Ben Stacy Jerrik (Ed.)

Sportavex

Ben Stacy Jerrik (Ed.)

Sportavex

Airshows, Contraction, Sport, Aviation, Exposition, New Zealand

Part Press

Imprint

Permission is granted to copy, distribute and/or modify this document under the terms of the GNU Free Documentation License, Version 1.2 or any later version published by the Free Software Foundation; with no Invariant Sections, with the Front-Cover Texts, and with the Back- Cover Texts. A copy of the license is included in the section entitled "GNU Free Documentation License".

All parts of this book are extracted from Wikipedia, the free encyclopedia (www.wikipedia.org).

You can get detailed informations about the authors of this collection of articles at the end of this book. The editors (Ed.) of this book are no authors. They have not modified or extended the original texts.

Pictures published in this book can be under different licences than the GNU Free Documentation License. You can get detailed informations about the authors and licences of pictures at the end of this book.

The content of this book was generated collaboratively by volunteers. Please be advised that nothing found here has necessarily been reviewed by people with the expertise required to provide you with complete, accurate or reliable information. Some information in this book maybe misleading or wrong. The Publisher does not guarantee the validity of the information found here. If you need specific advice (f.e. in fields of medical, legal, financial, or risk management questions) please contact a professional who is licensed or knowledgeable in that area.

Any brand names and product names mentioned in this book are subject to trademark, brand or patent protection and are trademarks or registered trademarks of their respective holders. The use of brand names, product names, common names, trade names, product descriptions etc. even without a particular marking in this works is in no way to be construed to mean that such names may be regarded as unrestricted in respect of trademark and brand protection legislation and could thus be used by anyone.

Cover image: www.ingimage.com
Concerning the licence of the cover image please contact ingimage.

Publisher:
Part Press is a trademark of
International Book Market Service Ltd., 17 Rue Meldrum, Beau Bassin, 1713-01 Mauritius
Email: info@bookmarketservice.com
Website: www.bookmarketservice.com

Published in 2012

Printed in: U.S.A., U.K., Germany. This book was not produced in Mauritius.

ISBN: 978-613-8-79105-8

Contents

Sportavex

SportAvex is a contraction of **Sport Av**iation **Ex**position. It is the annual fly-in and airshow organised by the Sport Aircraft Association New Zealand Incorporated (SAANZ).

This biennial event is the opportunity for homebuilders and other sport flyers to gather together and enjoy the spirit of camaraderie and competition with an aviation flavour. The event is held in alternate years with a similar but smaller event (Great Plains Flyin) held in the South Island. Each year, the top homebuilt aircraft completed in the preceding year are judged and champions are named and honoured. An airshow is conducted as part of the weekend's activities, featuring acts drawn from both the homebuilt/sport aviation and military aviation communities.

List of Venue/Years

- 1994 - Paraparaumu
- 1996 -
- 1998 - Matamata
- 2000 - Matamata
- 2002 - Matamata
- 2004 - Tauranga
- 2006 - Tauranga

External links

- The SAANZ website [1]
- The Tauranga Airshow Site [2]

References

[1] http://www.saa.org.nz/
[2] http://www.taurangaairshow.co.nz/

Homebuilt_aircraft

Homebuilt aircraft

A Rutan Long-EZ homebuilt in 1984 in England

Also known as *amateur-built aircraft* or *kit planes*, **homebuilt aircraft** are constructed by persons for whom this is not a professional activity. These aircraft may be constructed from "scratch," from plans, or from assembly kits.[1] [2]

Overview

In the United States, Brazil, Australia and New Zealand, homebuilt aircraft may be licensed Experimental under FAA or similar local regulations. Provided that the owner has done at least 51% of the construction work themselves they can also apply for a repairman's certificate for that airframe. The repairman's certificate allows the holder to perform and sign off on most of the maintenance, repairs, and inspections themselves.[1] [2]

Alberto Santos-Dumont was the first to offer for free construction plans, publishing drawings of its Demoiselle in the June 1910 edition of Popular Mechanics. The first aircraft to be offered for sale as plans, rather than a completed airframe, was the Baby Ace in the late 1920s.

Homebuilt aircraft gained in popularity in the US in 1924 with the start of the National Air Races, held in Dayton, Ohio. These races required aircraft with useful loads of 150 lb (68 kg) and engines of 80 cubic inches or less and as a consequence of the class limitations most were amateur-built. The years after Charles Lindbergh's transatlantic flight brought a peak of interest between 1929-33. During this period many aircraft designers, builders and pilots were self taught and the high accident rate brought public condemnation and increasing regulation to amateur-building. The resulting federal standards on design, engineering, stress analysis, use of aircraft-quality hardware and testing of aircraft brought an end to amateur building except in some specialized areas, such as racing. In 1946 Goodyear restarted the

Canada's first homebuilt aircraft, Stitts SA-3A Playboy CF-RAD, first flown in 1955, seen in the Canada Aviation and Space Museum.

National Air Races, including a class for aircraft powered by 200 cubic inch and smaller engines. The midget racer class spread nationally in the US and this led to calls for acceptable standards to allow recreational use of amateur-built aircraft. By the mid-1950s both the US and Canada once again allowed amateur-built aircraft to specified standards and limitations.[2]

Homebuilt aircraft are generally small, one to four-seat sportsplanes which employ simple methods of construction. Fabric-covered wood or metal frames and plywood are common in the aircraft structure, but increasingly, fiberglass and other composites as well as full aluminum construction techniques are being used, first pioneered by Hugo Junkers as far back as the late World War I era. Engines are most often the same as, or similar to, the engines used in certified aircraft (such as Lycoming, Continental, Rotax, and Jabiru). A minority of homebuilts use converted automobile engines, with Volkswagen air-cooled flat-4s, Subaru-based liquid-cooled engines, Mazda Wankel and Chevrolet Corvair six-cylinder engines being common. The use of automotive engines helps to reduce costs, but many builders prefer dedicated aircraft engines, which are perceived to have better performance and reliability. Other engines that have been used include chainsaw and motorcycle engines.[1] [2]

A combination of cost and litigation, especially in the mid-1980s era, which has discouraged general aviation manufacturers from introducing new designs, has led to homebuilts outselling factory types by five to one. In 2003, the number of homebuilts produced in the USA exceeded the number produced by any single certified manufacturer.

History

The history of amateur-built aircraft can be traced to the beginning of aviation. Even if the Wright brothers, Clément Ader, and their successors had commercial objectives in mind, the first aircraft were constructed by passionate enthusiasts whose goal was to fly.

Early years

Aviation took a leap forward with the industrialization that accompanied World War I. In the post-war period, manufacturers needed to find new markets and introduced models designed for tourism. However, these machines were affordable only by the very rich.

Many U.S. aircraft designed and registered in the 1920s onward were considered "experimental" by the (then) CAA, the same registration under which modern homebuilts are issued Special Airworthiness Certificates. Many of these were prototypes, but designs such as Bernard Pietenpol's first 1923 design were some of the first homebuilt aircraft. In 1928, Henri Mignet published plans for his HM-8 *Pou-du-Ciel*, as did Pietenpol for his Air Camper. Pietenpol later constructed a factory, and in 1933 began creating and selling partially-constructed aircraft kits.[2]

In 1936, an association of amateur aviation enthusiasts was created in France. Many types of amateur aircraft began to make an appearance, and in 1938 legislation was amended to provide for a *Certificat de navigabilité restreint d'aéronef* (*CNRA*, "restricted operating certificate for aircraft"). 1946 saw the birth of the Ultralight Aircraft Association which in 1952 became the Popular Flying Association in the United Kingdom, followed in 1953 by the Experimental Aircraft Association in the United States and the Sport Aircraft Association in Australia.

Technology and innovation

Until the late 1950s, builders had mainly kept to wood-and-cloth and steel tube-and-cloth design. Without the regulatory restrictions faced by production aircraft manufacturers, homebuilders introduced innovative designs and construction techniques. Burt Rutan introduced the canard design to the homebuilding world and pioneered the use of composite construction. Metal construction in kitplanes was taken to a new level by Richard VanGrunsven in his RV series. As the sophistication of the kits improved, components such as autopilots and more advanced navigation instruments became common.[1] [2]

The Questair Venture set new standards for speed in kit-built aircraft design

Litigation during the 1970s and 1980s caused stagnation in the small aircraft market, forcing the surviving companies to retain older, proven designs. In recent years, the less restrictive regulations for homebuilts allowed a number of manufacturers to develop new and innovative designs; many can outperform certified production aircraft in their class.

An example of high-end homebuilt design is Lancair, which has developed a number of high-performance kits. The most powerful is the Lancair Propjet, a four-place kit with cabin pressurization and a turboprop engine, cruising at 24000 feet (7300 m) and 370 knots (425 mph, 685 km/h). Although aircraft such as this are considered "home-built" for legal reasons, they are typically built in the factory with the assistance of the buyer. This allows the company which sells the kit to avoid the long and expensive process of certification, because they remain owner-built according to the regulations. One of the terms applied to this concept is commonly referred to as "The 51% Rule", which requires that builders perform the majority of the fabrication and assembly to be issued a Certificate of Airworthiness as an Amateur Built aircraft.[3]

A small number of jet kitplanes have been built since the 1970s, including the tiny Bede Aircraft BD-5J.[2]

Building materials

Homebuilt aircraft can be constructed out of any material that is light and strong enough for flight. Several common construction methods are detailed below.

Swearingen SX-300

Wood and fabric

This is the oldest construction, seen in the first aircraft and hence the best known. For that reason, amateur-built aircraft associations will have more specialists for this type of craft than other kinds.[1] [2]

The most commonly-used woods are Sitka spruce and Douglas fir, which offer excellent strength-to-weight ratios. Wooden structural members are joined with adhesive, usually epoxy. Unlike the wood construction techniques used in other applications, virtually all wooden joints in aircraft are simple butt joints, with plywood gussets. Joints are designed to be stronger than the members. After the structure has been completed, the aircraft is covered in aircraft fabric (usually

A typical wood and fabric construction amateur-built, the Bowers Fly Baby.

aircraft-grade polyester). The advantage of this type of construction is that it does not require complex tools and equipment, but commonplace items such as saw, planer, file, sandpaper, and clamps.[1] [2]

Examples of amateur-built wood and fabric designs include:

- The classic Pietenpol Air Camper, a homebuilt that has been built since the 1920s.
- The Bowers Fly Baby, a low-wing monoplane which has been popular since the 1960s.
- The Ison miniMAX

A Pietenpol Air Camper under construction, showing the wooden frame structure that will be covered with aircraft fabric.

Wood/composite mixture

A recent trend is toward wood-composite aircraft. The basic load carrying material is still wood, but it is combined with foam (for instance to increase buckling resistance of load carrying plywood skins) and other synthetic materials like glass- and carbon fibre (to locally increase the modulus of load carrying structures like spar caps, etc.).[1] [2]

Examples of wood-composite designs include:

- Ibis experimental aircraft project, designed by Roger Junqua
- KR series of homebuilts designed by Ken Rand
- PIK-26 designed by Kai Mellen

Metal

Planes built from metal use similar techniques to more conventional factory-built aircraft. They can be more challenging to build, requiring metal-cutting, metal-shaping, and riveting if building from plans. "Quick-build" kits are available which have the cutting, shaping and hole-drilling mostly done, requiring only finishing and assembly. Such kits are also available for the other types of aircraft construction, especially composite.[1] [2]

Van's Aircraft like this RV-4 are the most common metal homebuilt type.

There are three main types of metal construction: sheet aluminium, tube aluminium, and welded steel tube. The tube structures are covered in aircraft fabric, much like wooden aircraft.

Examples of metal-based amateur aircraft include:

- The Murphy Moose, Rebel, Super Rebel and Maverick, produced by Murphy Aircraft.
- The Vans RV-4, RV-8, RV-10 and other models produced by Van's Aircraft, are the most popular metal homebuilt aircraft.
- Chris Heintz's Zenith CH601 Zodiac and Zenith STOL CH701

Inside of the tail cone of a Murphy Moose under construction, showing the all-metal semi-monocoque design

Composite

Composite material structures are made of cloth with a high tensile strength (usually fiberglass or carbon fiber, or occasionally Kevlar) combined with a structural plastic (usually epoxy, although vinylester is used in some aircraft). The fabric is saturated with the structural plastic in a liquid form; when the plastic cures and hardens, the part will hold its shape while possessing the strength characteristics of the fabric.[1] [2]

The two primary types of composite planes are moulded composite, where major structures like wing skins and fuselage halves are prepared and cured in moulds, and mouldless, where shapes are carved out of foam and then covered with fiberglass or carbon fiber.[1] [2]

The advantages of this type of construction include smooth surfaces (without the drag of rivets), the ability to do compound curves, and the ability to place fiberglass or carbon fiber in optimal positions, orientations, and quantities. Drawbacks include the need to work with chemical products as well as low strength in directions perpendicular to fiber. Composites provide superb strength to weight. Material stiffness dependent upon direction (as opposed to equal in all directions, as with metals) allows for advanced "elastic tailoring" of composite parts.[1] [2]

A fiberglass/foam Quickie Q2.

A composite construction Cirrus VK-30.

Examples of amateur craft made of composite materials include:

- Canard designs such as the VariEze and Long EZ designed by Burt Rutan
- The pusher propeller Cirrus VK-30
- The Europa XS

Safety

The safety record of homebuilts is not as good as certified general aviation aircraft. In the United States, in 2003, amateur-built aircraft experienced a rate of 21.6 accidents per 100,000 flight hours; the overall general aviation accident rate for that year was 6.75 per 100,000 flight hours.[4]

The accident rate for homebuilt aircraft in the USA has long been a concern to the Federal Aviation Administration. At Sun 'n Fun 2010 FAA Administrator Randy Babbitt said that homebuilts "account for 10 percent of the GA fleet, but 27 percent of accidents. It's not the builders [getting into accidents], but the second owners. We need better transition training."[5]

Most nations' aviation regulations require amateur-built aircraft to be physically marked as such (for example in the UK "Occupant Warning - This aircraft ... is amateur built." must be displayed[6]), and extra flight testing is usually required before passengers (who are not pilots themselves) can be carried.

Culture

The largest airshow in the world is the Experimental Aircraft Association's annual EAA AirVenture Oshkosh airshow in Oshkosh, Wisconsin, which takes place in late July and early August. Other annual events are the Sun N' Fun Fly-In, which occurs in the early spring in Lakeland, Florida, and the Northwest EAA Fly-In in Arlington, Washington. These events are called a fly-in as many people fly their homebuilts and other aircraft into the airport hosting the show, often camping there for the duration. Both events last a week. Takeoffs and landings at these shows number in the thousands.

See also

- Aircraft design
- Ultralight
- Special Airworthiness Certificate

References

[1] Armstrong, Kenneth: *Choosing Your Homebuilt - the one you will finish and fly! Second Edition*, pages 39-52. Butterfield Press, 1993. ISBN 0-932579-26-4

[2] Peter M Bowers: *Guide to Homebuilts - Ninth Edition*. TAB Books, Blue Ridge Summit PA, 1984. ISBN 0-8306-2364-7

[3] Amateur-Built Aircraft - Federal Aviation Administration (http://www.faa.gov/aircraft/gen_av/ultralights/amateur_built/)

[4] National Transportation Safety Board (2007). "U.S. General Aviation, Calendar Year 2003" (http://www.ntsb.gov/publictn/2007/ ARG0701.pdf). *Annual Review of Aircraft Accident Data*. . Retrieved 2008-06-01.

[5] Grady, Mary (April 2010). "FAA Administrator Babbitt Takes In Sun 'n Fun" (http://www.avweb.com/news/snf/ SunNFun2010_FAAAdministratorBabbittTakesInSunnFun_202379-1.html). . Retrieved 17 April 2010.

[6] "CAP 659: Amateur Built Aircraft: A Guide to Approval, Construction, and Operation of Amateur Built Aircraft" (http://www.caa.co.uk/ docs/33/CAP659.PDF). Civil Aviation Authority. November 2005. .

External links

- Experimental Aircraft Association (http://www.eaa.org) (EAA)
- Light Aircraft Association (http://www.lightaircraftassociation.co.uk/), the representative body in the United Kingdom for amateur aircraft.
- FAA Advisory Circular 20-27G: Certification and Operation of Amateur-Built Aircraft (http://www.faa.gov/ documentLibrary/media/Advisory_Circular/AC 20-27G.pdf)
- FAA Advisory Circular 20-27F: Certification and Operation of Amateur-Built Aircraft (http://rgl.faa.gov/ Regulatory_and_Guidance_Library/rgAdvisoryCircular.nsf/0/0ca2845e2aafffbb86256dbf00640cb2/)

Aviation

<table>
<tr><td align="center">Aviation</td></tr>
<tr><td align="center"></td></tr>
<tr><td align="center">NASA Gulfstream V C-37A</td></tr>
</table>

Aviation is the design, development, production, operation, and use of aircraft, especially heavier-than-air aircraft. *Aviation* is derived from *avis*, the Latin word for *bird*.

History

Many cultures have built devices that travel through the air, from the earliest projectiles such as stones and spears,[1] [2] the boomerang in Australia, the hot air Kongming lantern, and kites. There are early legends of human flight such as the story of Icarus, and Jamshid in Persian myth, and later, somewhat more credible claims of short-distance human flights appear, such as the flying automaton of Archytas of Tarentum (428–347 BC),[3] the winged flights of Abbas Ibn Firnas (810–887), Eilmer of Malmesbury (11th century), and the hot-air Passarola of Bartolomeu Lourenço de Gusmão (1685–1724).

The modern age of aviation began with the first untethered human lighter-than-air flight on November 21, 1783, in a hot air balloon designed by the Montgolfier brothers. The practicality of balloons was limited because they could only travel downwind. It was immediately recognized that a steerable, or dirigible, balloon was required. Jean-Pierre Blanchard flew the first human-powered dirigible in 1784 and crossed the English Channel in one in 1785.

In 1799 Sir George Cayley set forth the concept of the modern airplane as a fixed-wing flying machine with separate systems for lift, propulsion, and control.[4] [5] Early dirigible developments included machine-powered propulsion (Henri Giffard, 1852), rigid frames (David Schwarz, 1896), and improved speed and maneuverability (Alberto Santos-Dumont, 1901)

First assisted take-off flight by the Wright Brothers, December 17, 1903

While there are many competing claims for the earliest powered, heavier-than-air flight, the most widely-accepted date is December 17, 1903 by the Wright brothers. The Wright brothers were the first to fly in a powered and controlled aircraft. Previous flights were gliders (control but no power) or free flight (power but no control), but the Wright brothers combined both, setting the new standard in aviation records. Following this, the widespread adoption of ailerons versus wing warping made aircraft much easier to control, and only a decade later, at the start of World War I, heavier-than-air powered aircraft had become practical for reconnaissance, artillery spotting, and even attacks against ground positions.

Aircraft began to transport people and cargo as designs grew larger and more reliable. In contrast to small non-rigid blimps, giant rigid airships became the first aircraft to transport passengers and cargo over great distances. The best known aircraft of this type were manufactured by the German Zeppelin company.

The most successful Zeppelin was the Graf Zeppelin. It flew over one million miles, including an around-the-world flight in August 1929. However, the dominance of the Zeppelins over the airplanes of that period, which had a range of only a few hundred miles, was diminishing as airplane design advanced. The "Golden Age" of the airships ended on May 6, 1937 when the Hindenburg caught fire,

Hindenburg at Lakehurst Naval Air Station, 1936

killing 36 people. Although there have been periodic initiatives to revive their use, airships have seen only niche application since that time.

Great progress was made in the field of aviation during the 1920s and 1930s, such as Charles Lindbergh's solo transatlantic flight in 1927, and Charles Kingsford Smith's transpacific flight the following year. One of the most successful designs of this period was the Douglas DC-3, which became the first airliner that was profitable carrying passengers exclusively, starting the modern era of passenger airline service. By the beginning of World War II, many towns and cities had built airports, and there were numerous qualified pilots available. The war brought many innovations to aviation, including the first jet aircraft and the first liquid-fueled rockets.

NASA's Helios researches solar powered flight.

After World War II, especially in North America, there was a boom in general aviation, both private and commercial, as thousands of pilots were released from military service and many inexpensive war-surplus transport and training aircraft became available. Manufacturers such as Cessna, Piper, and Beechcraft expanded production to provide light aircraft for the new middle-class market.

By the 1950s, the development of civil jets grew, beginning with the de Havilland Comet, though the first widely-used passenger jet was the Boeing 707, because it was much more economical than other planes at the time. At the same time, turboprop propulsion began to appear for smaller commuter planes, making it possible to serve small-volume routes in a much wider range of weather conditions.

Since the 1960s, composite airframes and quieter, more efficient engines have become available, and Concorde provided supersonic passenger service for more than two decades, but the most important lasting innovations have taken place in instrumentation and control. The arrival of solid-state electronics, the Global Positioning System, satellite communications, and increasingly small and powerful computers and LED displays, have dramatically changed the cockpits of airliners and, increasingly, of smaller aircraft as well. Pilots can navigate much more accurately and view terrain, obstructions, and other nearby aircraft on a map or through synthetic vision, even at night or in low visibility.

On June 21, 2004, SpaceShipOne became the first privately funded aircraft to make a spaceflight, opening the possibility of an aviation market capable of leaving the Earth's atmosphere. Meanwhile, flying prototypes of aircraft powered by alternative fuels, such as ethanol, electricity, and even solar energy, are becoming more common.

Civil aviation

Civil aviation includes all non-military flying, both general aviation and scheduled air transport.

Air transport

There are five major manufacturers of civil transport aircraft (in alphabetical order):

Northwest Airlines Airbus A330-323X

- Airbus, based in Europe
- Boeing, based in the United States
- Bombardier, based in Canada
- Embraer, based in Brazil
- United Aircraft Corporation, based in Russia

Boeing, Airbus, Ilyushin and Tupolev concentrate on wide-body and narrow-body jet airliners, while Bombardier, Embraer and Sukhoi concentrate on regional airliners. Large networks of specialized parts suppliers from around the world support these manufacturers, who sometimes provide only the initial design and final assembly in their own plants. The Chinese ACAC consortium will also soon enter the civil transport market with its ACAC ARJ21 regional jet.[6]

Until the 1970s, most major airlines were flag carriers, sponsored by their governments and heavily protected from competition. Since then, open skies agreements have resulted in increased competition and choice for consumers, coupled with falling prices for airlines. The combination of high fuel prices, low fares, high salaries, and crises such as the September 11, 2001 attacks and the SARS epidemic have driven many older airlines to government-bailouts, bankruptcy or mergers. At the same time, low-cost carriers such as Ryanair, Southwest and Westjet have flourished.

General aviation

General aviation includes all non-scheduled civil flying, both private and commercial. General aviation may include business flights, air charter, private aviation, flight training, ballooning, parachuting, gliding, hang gliding, aerial photography, foot-launched powered hang gliders, air ambulance, crop dusting, charter flights, traffic reporting, police air patrols and forest fire fighting.

1947 Cessna 120

Each country regulates aviation differently, but general aviation usually falls under different regulations depending on whether it is private or commercial and on the type of equipment involved.

Many small aircraft manufacturers serve the general aviation market, with a focus on private aviation and flight training.

The most important recent developments for small aircraft (which form the bulk of the GA fleet) have been the introduction of advanced avionics (including GPS) that were formerly found only in large airliners, and the introduction of composite materials to make small aircraft lighter and faster. Ultralight and homebuilt aircraft have also become increasingly popular for recreational use, since in most countries that allow private aviation, they are much less expensive and less heavily regulated than certified aircraft.

A weight-shift ultralight aircraft, the Air Creation Tanarg

Military aviation

Simple balloons were used as surveillance aircraft as early as the 18th century. Over the years, military aircraft have been built to meet ever increasing capability requirements. Manufacturers of military aircraft compete for contracts to supply their government's arsenal. Aircraft are selected based on factors like cost, performance, and the speed of production.

Types of military aviation

The Lockheed SR-71 remains unsurpassed in many areas of performance.

- Fighter aircraft's primary function is to destroy other aircraft. (e.g. Sopwith Camel, A6M Zero, F-15, MiG-29, Su-27, and F-22).
- Ground attack aircraft are used against tactical earth-bound targets. (e.g. Junkers Stuka, A-10, Il-2, J-22 Orao, AH-64 and Su-25).
- Bombers are generally used against more strategic targets, such as factories and oil fields. (e.g. Zeppelin, Tu-95, Mirage IV, and B-52).
- Transport aircraft are used to transport hardware and personnel. (e.g. C-17 Globemaster III, C-130 Hercules and Mil Mi-26).
- Surveillance and reconnaissance aircraft obtain information about enemy forces. (e.g. Rumpler Taube, Mosquito, U-2, OH-58 and MiG-25R).
- Unmanned aerial vehicles (UAVs) are used primarily as reconnaissance fixed-wing aircraft, though many also carry payloads. Cargo aircraft are in development. (e.g. RQ-7B Shadow, MQ-8 Fire Scout, and MQ-1C Gray Eagle).
- Missiles deliver warheads, normally explosives, but also things like leaflets.

Air Traffic Control (ATC)

Air traffic control towers at Amsterdam Airport

Air traffic control (ATC) involves communication with aircraft to help maintain *separation* — that is, they ensure that aircraft are sufficiently far enough apart horizontally or vertically for no risk of collision. Controllers may co-ordinate position reports provided by pilots, or in high traffic areas (such as the United States) they may use radar to see aircraft positions.

There are generally four different types of ATC:

- center controllers, who control aircraft en route between airports
- control towers (including tower, ground control, clearance delivery, and other services), which control aircraft within a small distance (typically 10–15 km horizontal, and 1,000 m vertical) of an airport.
- oceanic controllers, who control aircraft over international waters between continents, generally without radar service.
- terminal controllers, who control aircraft in a wider area (typically 50–80 km) around busy airports.

ATC is especially important for aircraft flying under Instrument flight rules (IFR), where they may be in weather conditions that do not allow the pilots to see other aircraft. However, in very high-traffic areas, especially near major airports, aircraft flying under Visual flight rules (VFR) are also required to follow instructions from ATC.

In addition to separation from other aircraft, ATC may provide weather advisories, terrain separation, navigation assistance, and other services to pilots, depending on their workload.

ATC do not control all flights. The majority of VFR flights in North America are not required to talk to ATC (unless they are passing through a busy terminal area or using a major airport), and in many areas, such as northern Canada and low altitude in northern Scotland, ATC services are not available even for IFR flights at lower altitudes.

Environmental impact

Like all activities involving combustion, operating powered aircraft (from airliners to hot air balloons) release soot and other pollutants into the atmosphere. Greenhouse gases such as carbon dioxide (CO_2) are also produced. In addition, there are environmental impacts specific to aviation:

- Aircraft operating at high altitudes near the tropopause (mainly large jet airliners) emit aerosols and leave contrails, both of which can increase cirrus cloud formation — cloud cover may have increased by up to 0.2% since the birth of aviation.[7]

- Aircraft operating at high altitudes near the tropopause can also release chemicals that interact with greenhouse gases at those altitudes, particularly nitrogen compounds, which interact with ozone, increasing ozone concentrations.[8] [9]

- Most light piston aircraft burn avgas, which contains tetra-ethyl lead (TEL). Some lower-compression piston engines can operate on unleaded mogas, and turbine engines and diesel engines — neither of which requires lead — are appearing on some newer light aircraft.

Water vapor contrails left by high-altitude jet airliners. These may contribute to cirrus cloud formation.

See also

- Aeronautics
- Aviation, aerospace, and aeronautical terms
- Environmental impact of aviation
- List of aviation topics
- Timeline of aviation

Notes

[1] Archytas of Tar entum, Technology Museum of Thessaloniki, Macedonia, Greece (http://www.tmth.edu.gr/en/aet/1/14.html)

[2] Automata history (http://automata.co.uk/History page.htm)

[3] " *Aviation: Reaching for the Sky* (http://books.google.com/books?id=Efr2Ll1OdqMC&pg=PA128&dq&hl=en#v=onepage&q=&f=false)". Don Berliner (1996). The Oliver Press, Inc. p.28. ISBN 1881508331

[4] "Aviation History" (http://www.aviation-history.com/early/cayley.htm). . Retrieved 2009-07-26.

[5] "Sir George Carley (British Inventor and Scientist)" (http://www.britannica.com/EBchecked/topic/100795/Sir-George-Cayley-6th-Baronet). Britannica. . Retrieved 2009-07-26. "English pioneer of aerial navigation and aeronautical engineering and designer of the first successful glider to carry a human being aloft."

[6] Kingsbury, Kathleen (October 11, 2007). "Eyes on the Skies" (http://www.time.com/time/magazine/article/0,9171,1670256,00.html). *Time*. . Retrieved April 26, 2010.

[7] Aviation and the Global Atmosphere (IPCC) (http://www.grida.no/climate/ipcc/aviation/032.htm)

[8] Lin, X.; Trainer, M. and Liu, S.C., (1988). "On the nonlinearity of the tropospheric ozone production.". *Journal of Geophysical Research* **93** (D12): 15879–15888. Bibcode 1988JGR....9315879L. doi:10.1029/JD093iD12p15879.

[9] Grewe, V.; D. Brunner, M. Dameris, J. L. Grenfell, R. Hein, D. Shindell, J. Staehelin (July 2001). "Origin and variability of upper tropospheric nitrogen oxides and ozone at northern mid-latitudes" (http://linkinghub.elsevier.com/retrieve/pii/S1352231001001340). *Atmospheric Environment* **35** (20): 3421–3433. doi:10.1016/S1352-2310(01)00134-0. . Retrieved 2007-11-20.

ltg:Aviaceja

Air_show

An **air show** is an event at which aviators display their flying skills and the capabilities of their aircraft to spectators in aerobatics. Air shows without aerobatic displays, having only aircraft displayed parked on the ground, are called "static air shows".

The UK Utterly Butterly display team flying Boeing Stearman PT-17 biplanes at an English air show

Outline

Some air shows are held as a business venture or as a trade event where aircraft, avionics and other services are promoted to potential customers. Many air shows are held in support of local, national or military charities. Military air firms often organise air shows at military airfields as a public relations exercise to thank the local community, promote military careers and raise the profile of the military.

A month after Blériot's crossing of the English Channel the aviation week in Reims, France, August 1909, caught special worldwide attention.

Aviation Nation 2006 at Nellis Air Force Base, United States

Air show "seasons" vary around the world. Whereas the United States enjoys a long season that generally runs from March to November, other areas often have much shorter seasons. The European season usually starts in late April or Early May and is usually over by mid October. The Middle East, Australia and New Zealand hold their events between January and March. However, for many acts the "off season" does not mean a period of inactivity, they use time for maintenance and practice.

The type of displays seen at an event are constrained by a number of factors, including the weather and visibility. Most aviation authorities now publish rules and guidance on minimum display heights and criteria for differing conditions. In addition to the weather, pilots and organizers must also consider local airspace restrictions. Most exhibitors will plan "full," "rolling" and "flat" display for varying weather and airspace conditions.

The Patrouille Suisse performing at the ILA Berlin Air Show

The types of shows vary greatly. Some are large scale military events with large flying displays and ground exhibitions while others held at small local airstrips can often feature just one or two hours of flying with just a few stalls on the ground. Air Displays can be held during day or night with the latter becoming increasingly popular. Shows don't always take place over airfields; some have been held over the grounds of stately homes or castles and over the sea at coastal resorts.

Attractions

Air racing at an air show in England: the Red Bull Air Race heat held at Kemble Airfield, Gloucestershire. The aircraft fly singly, and pass between pairs of pylons

Before the Second World War, air shows were associated with long distance air races, often lasting many days and covering thousands of miles. While the Reno Air Races keep this tradition alive, most air shows today primarily feature a series of aerial demos of short duration.

Most air shows will feature warbirds, aerobatics, and demonstrations of modern military aircraft, and many air shows offer a variety of other aeronautical attractions as well, such as wing-walking, radio-controlled aircraft, water/slurry drops from firefighting aircraft, simulated helicopter rescues and sky diving.

Specialist aerobatic aircraft have powerful piston engines, light weight and big control surfaces, making them capable of very high roll rates and accelerations. A skilled pilot will be able to climb vertically, perform very tight turns, tumble his aircraft end-over-end and perform manoeuvres during loops.

Solo military jet demos, also known as tactical demo, feature one aircraft, usually a strike fighter or an advanced trainer. The demonstration focuses on the capabilities of modern aircraft used in combat operations. The display will usually demonstrate the aircraft's very short (and often very loud) takeoff rolls, fast speeds, slow approach speeds, as well as their ability to quickly make tight turns, to climb quickly, and their ability to be precisely controlled at a large range of speeds. Manoeuvres include aileron rolls, barrel rolls, hesitation rolls, Cuban-8s, tight turns, high-alpha flight, a high-speed pass, double Immelmans, and touch-and-gos. Tactical demos may

RAAF F-111 performing a dump-and-burn fuel dump at the Australian International Airshow

include simulated bomb drops, sometimes with pyrotechnics on the ground for effect. Aircraft with special characteristics that give them unique capabilities will often display those in their demos; For example, Russian fighters with Thrust vectoring may be used to perform Pugachev's Cobra or the Kulbit, among other difficult manoeuvers that cannot be performed by other aircraft. Similarly, an F-22 pilot may hover his jet in the air with the nose pointed straight up, a Harrier or Osprey pilot may perform a vertical landing or vertical takeoff, and so on.

Safety

Air shows present some risk to spectators and aviators. Accidents have occurred, sometimes with a large loss of life, such as the 1988 disaster at Ramstein Air Base in Germany and the 2002 air show crash at Lviv, Ukraine. Because of these accidents, the various aviation authorities around the world have created set rules and guidance for those running and participating in air displays. Air displays are often monitored by aviation authorities to ensure safe procedures.

Rules govern the distance from the crowds that aircraft must fly. These vary according to the rating of the pilot/crew, the type of aircraft and the way the aircraft is being flown. For instance, slower lighter aircraft are usually allowed closer and lower to the crowd than larger, faster types. Also, a fighter jet flying straight and level will be able to do so closer to the crowd and lower than if it were performing a roll or a loop.

Pilots can get authorisations for differing types of displays (i.e. limbo flying, basic aerobatics to unlimited aerobatics) and to differing minimum base heights above the ground. To gain such authorisations, the pilots will have to demonstrate to an examiner that they can perform to those limits without endangering themselves, ground crew or spectators.

Mountain Home Air Force Base, Idaho, September 14, 2003: U.S. Air Force Thunderbirds Captain Christopher Stricklin ejecting from his F-16 after realizing he could not pull up in time from a Split-S and ensuring the aircraft would not crash into spectators. The aircraft was destroyed less than a second later with no loss of life.

Despite display rules and guidances, accidents have continued to happen. However, air show accidents are rare and where there is proper supervision air shows have impressive safety records. Each year, organisations such as The International Council of Air Shows [1] and The European Airshow Council [2] meet and discuss various subjects including air show safety where accidents are discussed and lessons learnt.

Weather

Air shows, and other big shows such as agricultural shows, on grassy land, are vulnerable to continued heavy rain waterlogging the ground, and making the cloudbase too low for flying, forcing cancellation, or the show ending early, costing much money for the show's organizers, as people and parking cars have difficulty moving about and turn the land into a morass, and the organizers may be tempted to put straw or cinders down to make movement easier, and the owner of the land cannot accept the resulting damage.

See also

- Airshow pilot
- Fly-in
- Flypast
- List of airshow accidents
- List of airshows
- Whifferdill turn

References

[1] http://www.airshows.org
[2] http://www.european-airshow.com

External links

- Experimental Aircraft Association Calendar (http://www.eaa.org/calendar/)
- Royal Aero Club (http://www.royalaeroclub.org/events.htm)
- Upcoming air shows (http://www.flightglobal.com/air-shows/)

Paraparaumu

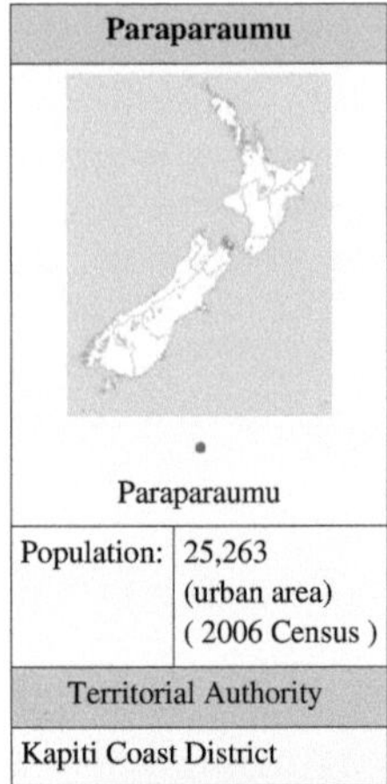

Paraparaumu	
Paraparaumu	
Population:	25,263 (urban area) (2006 Census)
Territorial Authority	
Kapiti Coast District	

Paraparaumu (Maori: ['paɾapa'raɔmʉ])[1] is a town in the south-western North Island of New Zealand. It lies in the Kapiti Coast, 50 kilometres north of the nation's capital city, Wellington.

Like other towns in the area, it has a partner settlement at the coast called Paraparaumu Beach, which lies directly opposite Kapiti Island. The two towns form part of the Kapiti Coast District. Together with the nearby Raumati Beach and Raumati South they are among the fastest-growing urban areas in New Zealand, and are major dormitory towns with workers commuting

Paraparaumu Airport

to the cities that make up the Wellington urban area. The four towns between them have a population of over 25,000 people.[2] Inland behind Paraparaumu is the Maungakotukutuku area.

Paraparaumu is home to the Kapiti Coast's largest secondary school, Paraparaumu College, with Kapiti College residing in nearby Raumati Beach and Otaki College in Otaki.

Paraparaumu was formerly represented in soccer by Paraparaumu United. They merged with the Raumati Hearts in 2003 to create Kapiti Coast United, which is based at Weka Park in Raumati.

Paraparaumu means "scraps from an earth oven" in Māori *parapara' means dirt or scraps and 'umu' means Oven.*[3] *It is commonly abbreviated to "Para-Param", particularly by longer-term residents of European ethnicity, and simply "Pram" by local youth.*

Born in Paraparaumu

* Christian Cullen – Rugby union footballer
* Cliff Curtis – Film actor
* John Henwood – Professional Runner, 2004 Olympian
* Stephen Kearney – Rugby league footballer and coach
* Wayne McIndoe – Field hockey player
* Drew Neemia – C4's Select Live host
* Andrew Niccol – Film director

Educated in Paraparaumu

* Peter Jackson at Kapiti College – Film Director
* Danny Page at Paraparaumu College - Professional Basketball Coach

Paraparaumu Airport

Paraparaumu Airport is a popular recreational airfield and hosts the Kapiti Aero Club. Apart from fixed wing and glider training, local helicopter operator – Helipro carries out extensive helicopter pilot training.

Public transport

Paraparaumu is located on the North Island Main Trunk Railway, on the Kapiti Line of Wellington's commuter railway network operated by Tranz Metro under the Metlink brand. Electrified commuter services were extended to Waikanae on 20 February 2011, and additional stations have been proposed at Lindale and Raumati. EM and FP class electric multiple units operate the commuter trains. Beyond Paraparaumu, Tranz Scenic operates two diesel-hauled long distance services: the Capital Connection between Palmerston North and Wellington, and the Overlander between Auckland and Wellington. Both trains stop at Paraparaumu. There are also feeder and local commuter bus services.

Notes

[1] English: /ˌpærəpəˈraʊmuː/ *parr-ə-pə-row-moo*, though typically pronounced English pronunciation: /ˌpærəˌpærəˈuːmuː/ *parr-ə-parr-ə-oo-moo*. In Maori vowels are run together, even when they are brought together by the creation of compound words. See for example (http://www.maorilanguage.info/mao_phon_desc1.html)

[2] Kapiti Coast Distinct Council, *Community Profile* (http://www.kapiticoast.govt.nz/NR/rdonlyres/ FEF38C62-2C46-47BA-929E-0E2207B08F91/39376/KapitiCoastCOMMUNITYPROFILE2006.pdf)

[3] Chris Maclean, "Wellington Places – Kapiti Coast", *Te Ara* (http://www.teara.govt.nz/Places/Wellington/WellingtonPlaces/15/en)

Matamata

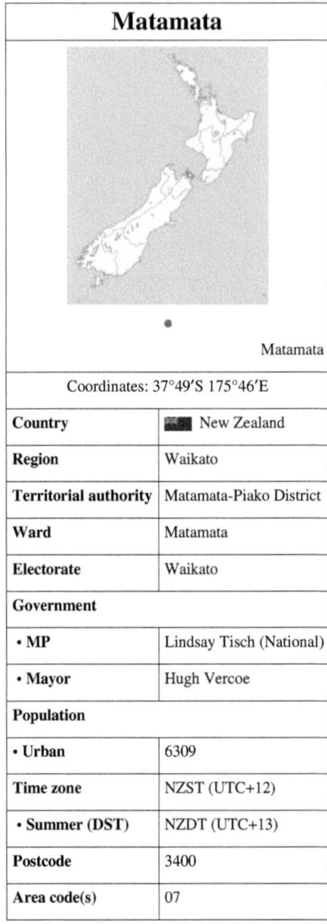

Matamata

Matamata

Coordinates: 37°49′S 175°46′E

Country	New Zealand
Region	Waikato
Territorial authority	Matamata-Piako District
Ward	Matamata
Electorate	Waikato
Government	
• **MP**	Lindsay Tisch (National)
• **Mayor**	Hugh Vercoe
Population	
• **Urban**	6309
Time zone	NZST (UTC+12)
• **Summer (DST)**	NZDT (UTC+13)
Postcode	3400
Area code(s)	07

Matamata is a rural Waikato town in New Zealand with a population of around 12,000 (6,000 in rural areas, 6,000 in the township). It is located near the base of the Kaimai Ranges, and is a thriving farming area known for Thoroughbred horse breeding and training pursuits. It is part of the Matamata-Piako District, which takes in the surrounding rural areas as well as Morrinsville and Te Aroha. State Highway 27 and the Kinleith Branch railway run through the town.

A nearby farm was the location for the Hobbiton set in Peter Jackson's *Lord of the Rings film trilogy*. The New Zealand government decided to leave the Hobbit holes built on location as tourist attractions, since they were designed to blend seamlessly into the environment. During

Sign in Matamata identifying town as the location where the Hobbiton scenes from the *Lord of the Rings* were filmed

the interim period filming Return of the King and The Hobbit they had no furniture or props, but could be entered with vistas of the farm viewed from inside them.[1] A "Welcome to Hobbiton" sign has been placed on the main road. In 2011 parts of Hobbiton began to close in preparation for 2 new movies based around the Lord of the Rings prequel novel The Hobbit.

Matamata is also the home of various media outlets, including studios for tvCentral, TV Rotorua and iTV Live, which is unusual for a town of such size.

Media

- tvCentral
- Scene
- Matamata Chronicle

Association football

Matamata is home to Matamata Swifts who compete in the Lotto Sport Italia NRFL Division 1A.

Schools

Secondary schools

- Matamata College

Intermediate schools

- Matamata Intermediate

Primary schools

- Hinuera School
- Firth Primary School
- Matamata Primary School

Religious schools

- Matamata Christian School
- St Josephs Catholic School

Notable people

- Dame Catherine Tizard (former Governor General of NZ)
- Shane Dye - jockey

Nearby towns

- Hinuera
- Peria
- Turanga-o-moana

- Te Poi
- Waharoa
- Walton
- Wardville

See also

- List of towns in New Zealand

References

[1] "Matamata Area Guide" (http://www.exploring.co.nz/North_Island/Waikato/Matamata/). . Retrieved 2009-01-07.

Tauranga

Tauranga	
Tauranga-moana (Māori)	
— Metropolitan Area —	

Tauranga

Location of Tauranga

Coordinates: 37°41′S 176°10′E

Country	New Zealand
Region	Bay of Plenty
Territorial authority	Tauranga City
Settled	1838
Gazetted as a Borough	1882
City constituted	17 April 1963
Electorate(s)	Tauranga Bay of Plenty
Government	
• MP (Tauranga)	Simon Bridges (National)
• MP (Bay of Plenty)	Tony Ryall (National)
• Mayor	Stuart Crosby

• Deputy Mayor	David Stewart
Area	
• Territorial	168 km^2 (64.9 sq mi)
Highest elevation	232 m (761 ft)
Lowest elevation	0 m (0 ft)
Population (June 2011 estimate)[1]	
• Territorial	115,700
• Density	**unknown operator: u','/km^2 (/sq mi)**
• Urban	121,500
Time zone	NZST (UTC+12)
• Summer (DST)	NZDT (UTC+13)
Postcode(s)	3110 - 3112 - 3113 - 3118
Area code(s)	07
Website	www.Tauranga.govt.nz [2]

Tauranga (🔊 /ˈtoʊ.rɑːŋɑː/ *TOW-ra-ngə*)[3] [4] is the most populous city in the Bay of Plenty region, in the North Island of New Zealand.

It was settled by Europeans in the early 19th century and was constituted as a city in 1963.[5] Tauranga City is the centre of the sixth largest urban area in New Zealand, with an urban population of 121,500 (June 2011 estimate).[1]

The city lies in the north-western corner of the Bay of Plenty, on the south-eastern edge of Tauranga Harbour. The city expands over an area of 168 square kilometres (65 sq mi), and encompasses the communities of Bethlehem, on the south-western outskirts of the city; Greerton, on the southern outskirts of the city; Matua, west of the central city overlooking Tauranga Harbour; Maungatapu; Mount Maunganui, located north of the central city across the harbour facing the Bay of Plenty; Otumoetai; Papamoa, Tauranga's largest suburb, located on the Bay of Plenty; Tauranga City; Tauranga South; and Welcome Bay.

Tauranga is one of New Zealand's main centres for business, international trade, culture, fashion and horticultural science. The Port of Tauranga is New Zealand's largest port in terms of gross export tonnage.[6] [7]

Tauranga is one of New Zealand's fastest growing cities, with a 14 percent increase in population between the 2001 census and the 2006 census.[8]

History

First settlers

The earliest known settlers arrived in Tauranga from the Takitimu and the Mataatua waka in the 12th century.[9]

Early trading

Traders in flax were active in the Bay of Plenty during the 1830s; some were transient, others married local women and settled permanently. The first permanent trader was James Farrow, who travelled to Tauranga in 1829, obtaining flax fibre for Australian merchants in exchange for muskets and gunpowder. Farrow acquired a land area of 0.5 acres (2000 m^2) on 10 January 1838 at Otumoetai Pā from the chiefs Tupaea, Tangimoana and Te Omanu, the earliest authenticated land purchase in the Bay of Plenty.[10]

During the 1820s, Henry Williams travelled to Tauranga from the Bay of Islands to obtain supplies of potatoes, pigs and flax. In 1835 a Church Missionary Society mission station was established at Tauranga by William Wade; Rev. A.N. Brown, arrived at the CMS mission station in 1838.[11]

In 1840, a Catholic mission station was established. Bishop Pompallier was given land within the palisades of Otumoetai Pā for a church and a presbytery. The mission station closed in 1863 due to land wars in the Waikato district.

View over Greater Tauranga, taken from the top of Mauao

New Zealand land wars

The Tauranga Campaign took place in and around Tauranga from 21 January to 21 June 1864, during the land wars. The Battle of Gate Pa is the best known.

Modern age

Under the *Local Government (Tauranga City Council) Order 2003*,[12] Tauranga became legally a city for a second time, from 1 March 2004.

In August 2011, Tauranga received "ultra fast" broadband as part of the New Zealand Government's rollout. [13]

Geography

Tauranga is located around a large harbour that extends along the western Bay of Plenty, and is protected by Matakana Island and the extinct volcano of Mauao.

Situated along a faultline, Tauranga and the Bay of Plenty experience infrequent seismic activity, and there are a few volcanoes around the area (mainly dormant). The most notable of these are White Island and Mauao (Mount Maunganui), nicknamed "The Mount" by locals.

Tauranga is roughly the antipodal point of Jaen, Spain.

Climate

Due to its sheltered position on the east coast, Tauranga enjoys a warm, subtropical climate.. During the summer months the population swells as the holidaymakers descend on the city, especially along the popular white coastal surf beaches from Mount Maunganui to Papamoa.

Population

Tauranga surpassed Dunedin in 2008 as sixth largest city in New Zealand by urban area, and ninth largest city by Territorial Authority area. The city was growing at a rate of 1.5% in 2008.

In 1976, Tauranga was a medium-sized urban area, with a population of around 48,000, smaller than Napier or Invercargill. The completion of a harbour bridge in 1988 brought Tauranga and The Mount closer (they amalgamated in 1989) and promoted growth in both parts of the enlarged city. In 1996 Tauranga's population was 82,092 and by 2006 it had reached 103,635.[15]

In 2006, 17.4% of the population was aged 65 or over, compared to 12.3% nationally. The city hosts five major head offices – Port of Tauranga, Zespri International, Ballance Agri-Nutrients Ltd, Trustpower and Craigs Investment Partners (formerly, ABN AMRO Craigs). Tauranga is home to a large number of migrants, especially from the UK, attracted to the area by its climate and quality of life.

Mount Maunganui Main Beach in winter, with 'Leisure Island' in the background.

Economy

Much of the countryside surrounding Tauranga is horticultural land, used to grow a wide range of fresh produce for both domestic consumption and export. The area is particularly well known for growing tangelos (a grapefruit / tangerine cross), avocados, and kiwifruit. Recent years have seen the establishment of boutique vineyards and wineries.

The Port of Tauranga is New Zealand's largest export port, with brisk but seasonal shipping traffic. It is a regular stop for both container ships and luxury cruise liners.

Tauranga's main shopping mall is Bayfair, in Mount Maunganui. Most of the city's shopping centres are located in the suburbs. They include Fraser Cove , Bethlehem Town Centre, Fashion Island, Bayfair Mall, Mount Maunganui Main Street, Bay Central and Greerton Village. In 2008 Tauranga's CBD underwent renovations to attract more shoppers to the inner city.

Tauranga harbour.

Arts and Culture

Religion

Because of Tauranga's large multi-ethnic population, a wide variety of faiths are practiced, including Islam, Buddhism, Hinduism, various Eastern Orthodox Churches, Sufism and others. Immigrants from Asia have formed a number of significant Buddhist congregations.

The main faith present throughout Tauranga and most of the country is Christianity. There are many types of Christians, including: the largest, Pentecostal, Methodist and Presbyterian. There are also congregations of Mormons and in the area, and Jews.

Attractions

Greater Tauranga is a very popular lifestyle and tourism destination. It features many natural attractions and scenery. Cultural attractions

Picturesque sunrise over the Tauranga harbour.

include the Tauranga Art Gallery [16], which opened in October 2007 and showcases local, national and international exhibitions in a range of media.

View of Mount Beach, with Mauao in background

Events

The National Jazz Festival [17] takes place in Tauranga every Easter, with dozens of live acts, great food and excellent wine.

New Year celebrations at the Mount in Mount Maunganui are one of Tauranga's main events, bringing people from all around the country.

Lifestyle

The coastal suburb Papamoa and neighbouring town Mount Maunganui are some of the more affluent areas around Tauranga. The region's beaches attract swimmers, surfers, kayakers and kitesurfers throughout the year.

Tauranga has many outlying islands and reefs that make it a notable tourist destination point for traveling scuba divers and marine enthusiasts. Extensive marine life diversity is available to scuba divers all year round. Water temperatures range from 12 degrees Celsius in winter to 22-24 degrees Celsius in summer. Tauranga houses two professional dive instructor training centers, training NAUI, PADI and SSI dive leader systems.

Sports

Tauranga has a large stadium complex in the Bayfair suburb, Baypark Stadium, rebuilt in 2001 after a similar complex closed in 1995. It hosts Speedway events during summer and rugby matches in winter.

Tauranga is also the home of Football (Soccer) club Tauranga City United who compete in the Lotto Sport Italia NRFL Division 2.

Parks and recreation

Tauranga has many parks. One of the largest is Memorial Park, and others include, Yatton Park, Kulim Park, Fergusson Park and the large Tauranga Domain.

Due to the temperate climate, outdoor activities are very popular, including golf, tramping (hiking), mountain biking and white water rafting. The Bay of Plenty coastline has miles of golden sandy beaches, and watersports are very popular, including swimming, surfing, fishing, diving, kayaking and kitesurfing. Tourists also enjoy dolphin-watching on specially run boat trips.

McLaren Falls Park, on the outskirts of Tauranga

Education

Tauranga is home to the Bay of Plenty Polytechnic and a branch of the University of Waikato.

The main state secondary schools include:

- Papamoa College [18], co-educational secondary school opened in 2011 for years 7 - 13.
- Aquinas College a co-educational state-integrated Catholic school founded in 2003 for years 7 - 13, with around 800 pupils.
- Tauranga Boys' College, with over 1800 boys.
- Tauranga Girls' College, with over 1500 girls.
- Otumoetai College, with around 1900 pupils.
- Bethlehem Campus, a Christian educational institution for kindergarten, primary and secondary level students, with around 1500 students.
- Mount Maunganui College, a co-educational secondary school, with over 1500 students.

Christian educational institutions in Tauranga include Bethlehem Campus, a college for both children and adults established in 1988; Tauranga Adventist School [19], a state integrated Christian community school catering for Year 1 to 8 students and established in 1974; and Aquinas College [20], a Catholic college established in 2003.

There is also a Rudolf Steiner School [21] in Welcome Bay, catering for birth to 12 year olds.

Transportation

Main transportation in the city is provided by the Bay Hopper Bus. Tauranga Airport provides daily domestic flights to many destinations in the country.

Notable residents

- Hilda Hewlett - pioneer aviator
- Phil Rudd - drummer for AC/DC
- Richard O'Brien - author of The Rocky Horror Show (spent his formative years here)
- Les Munro - Dambusters veteran.
- Bob Clarkson - former Member of Parliament and successful property developer and landlord
- Mahé Drysdale - Olympic rower
- Andrew (Herb) Stevenson - Olympic rower, Double World Champion Rower, NZ 1982 Sportsman of the Year
- Tim Balme - actor, director
- Moss Burmester - Olympic swimmer
- Simon Bridges - politician
- John Bracewell - International Cricketer
- Dame Susan Devoy - former World Squash Champion
- Tony Lochhead - footballer
- Kane Williamson - International Cricketer
- Tony Christiansen - Former Paralympics, FESPIC Games & World Games multi-medalist, Professional Speaker & Tauranga City Councillor

Past residents

- Winston Peters - former MP for Tauranga, leader of NZ First, politician
- Stan Walker - R&B singer, Former Australian Idol contestant and winner [22]
- Jeremy Redmore - Midnight Youth singer

Sister cities

- ● Hitachi, Ibaraki, Japan[23]
- Yantai, Shandong, China

References

[1] "Subnational population estimates at 30 June 2011 (boundaries at 1 July 2011)" (http://www.stats.govt.nz/browse_for_stats/population/ estimates_and_projections/subnational-pop-estimates-tables.aspx). Statistics New Zealand. 19 December 2011. . Retrieved 19 December 2011.

[2] http://www.tauranga.govt.nz/

[3] (http://www.nzhistory.net.nz/culture/tereo-100words)

[4] http://www.forvo.com/word/tauranga/

[5] "Local Government 1860 - present" (http://www.library.tauranga.govt.nz/localhistory/european-settlement/the-rule-of-law.aspx). Tauranga City Council. . Retrieved 2008-12-19.

[6] http://business.newzealand.com/common/files/New-Zealand-ports-and-airports.pdf

[7] http://www.port-tauranga.co.nz/

[8] Quickstats about Tauranga City (http://www.stats.govt.nz/Census/2006CensusHomePage/QuickStats/AboutAPlace/SnapShot. aspx?id=2000023)

[9] VTours - Downtown Tauranga (http://www.citynews.co.nz/vtours/tauranga/index.html)

[10] http://www.tauranga.govt.nz/knowledgebase/tabid/624/qid/1164/tctl/1332_ViewQuestion/Default.aspx

[11] Rogers, Lawrence M. (1973). *Te Wiremu: A Biography of Henry Williams*. Pegasus Press.

[12] http://www.knowledge-basket.co.nz/regs/regs/text/2003/2003275.txt

[13] http://www.voxy.co.nz/technology/ultra-fast-broadband-comes-tauranga/5/99115

[14] "Climate Data" (http://www.niwascience.co.nz/edu/resources/climate/). NIWA. . Retrieved 2 November 2007.

[15] Geonames Database "Region information page" (http://www.travelsradiate.com/oceania/new-zealand/tauranga/2208032), *travelsradiate*, 2010, accessed 2 January 2011.

[16] http://www.artgallery.org.nz/index.php/pi_pageid/2

[17] http://www.jazz.org.nz/

[18] http://www.papamoacollege.school.nz/

[19] http://www.taurangasda.school.nz/

[20] http://www.aquinas.school.nz/

[21] http://www.rudolfsteinertga.ac.nz/

[22] "Kiwi leads Idol show by a neck" (http://www.stuff.co.nz/entertainment/2849353/Kiwi-leads-Idol-show-by-a-neck/). The Dominion Post. 10 September 2009. . Retrieved 2009-11-03.

[23] Sister cities (http://www.tauranga.govt.nz/water/city-waters-projects/waiari-scheme/why/tabid/674/qid/21/tctl/ 1447_ViewQuestion/Default.aspx)

External links

- Tauranga City Council (http://www.tauranga.govt.nz/)]
- Tauranga Portal Website (http://www.tauranga.co.nz/)
- Tauranga Local Information website (http://www.cityoftauranga.co.nz/)

New_Zealand

<table>
<tr><td colspan="3" align="center">New Zealand
Aotearoa</td></tr>
<tr><td colspan="3" align="center"> </td></tr>
<tr><td colspan="3" align="center">Anthem:
"God Defend New Zealand"
"God Save the Queen"[1]</td></tr>
<tr><td colspan="3" align="center">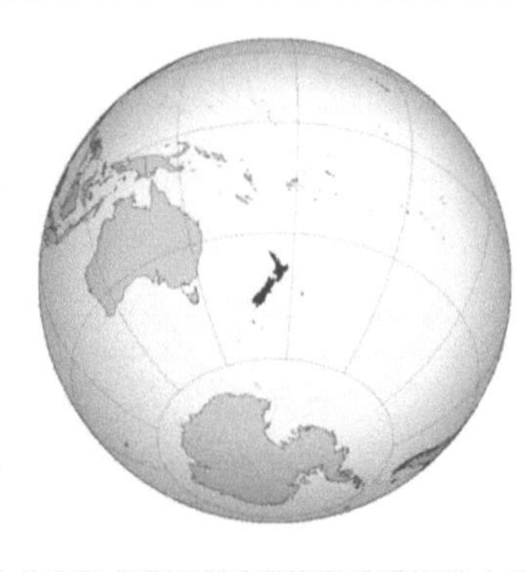
The hemisphere centred on New Zealand</td></tr>
<tr><td>Capital</td><td colspan="2">Wellington
41°17′S 174°27′E</td></tr>
<tr><td align="center">Largest city</td><td colspan="2">Auckland</td></tr>
<tr><td align="center">Official language(s)</td><td colspan="2">Māori (4.2%)[2]
NZ Sign Language (0.6%)</td></tr>
<tr><td align="center">National language</td><td colspan="2">English (98%)</td></tr>
<tr><td>Ethnic groups</td><td colspan="2">78% European/Other[3]
14.6% Māori
9.2% Asian
6.9% Pacific peoples</td></tr>
<tr><td align="center">Demonym</td><td colspan="2">New Zealander,
Kiwi (colloquial)</td></tr>
<tr><td align="center">Government</td><td colspan="2">Parliamentary constitutional monarchy</td></tr>
<tr><td>-</td><td>Monarch</td><td>Elizabeth II</td></tr>
<tr><td>-</td><td>Governor-General</td><td>Sir Jerry Mateparae</td></tr>
<tr><td>-</td><td>Prime Minister</td><td>John Key</td></tr>
<tr><td align="center">Independence</td><td colspan="2">from the United Kingdom[4]</td></tr>
</table>

-	1st Parliament	25 May 1854
-	Dominion	26 September 1907
-	Statute of Westminster	11 December 1931 (adopted 25 November 1947)
-	Constitution Act 1986	13 December 1986
Area		
-	Total	268,021 km^2 (75th) 103,483 sq mi
-	Water (%)	1.6[5]
Population		
-	September 2011 estimate	4,414,400[6] (124th)
-	2006 census	4,027,947[7]
-	Density	16.5/km^2 (202nd) 42.7/sq mi
GDP (PPP)		2010 estimate
-	Total	$117.807 billion[8] (61st)
-	Per capita	$26,966[8] (32nd)
GDP (nominal)		2010 estimate
-	Total	$140.434 billion[8] (51st)
-	Per capita	$32,145[8] (24th)
Gini (1997)		36.2[9] (medium)
HDI (2011)		▲ 0.908[10] (very high) (5th)
Currency		New Zealand dollar (NZD)
Time zone		NZST[11] (UTC+12)
-	Summer (DST)	NZDT (UTC+13)
		(Sep to Apr)
Date formats		dd/mm/yyyy
Drives on the		left
ISO 3166 code		NZ
Internet TLD		.nz[12]
Calling code		+64

New Zealand (Māori: *Aotearoa*) is an island country in the south-western Pacific Ocean comprising two main landmasses (the North Island and the South Island) and numerous smaller islands. The country is situated some 1500 kilometres (900 mi) east of Australia across the Tasman Sea, and roughly 1000 kilometres (600 mi) south of the Pacific island nations of New Caledonia, Fiji, and Tonga. Because of its remoteness, it was one of the last lands to be settled by humans. During its long isolation New Zealand developed a distinctive fauna dominated by birds, many of which became extinct after the arrival of humans and introduced mammals. With a mild maritime climate, the land was mostly covered in forest. The country's varied topography and its sharp mountain peaks owe much to the uplift of land and volcanic eruptions caused by the Pacific and Indo-Australian Plates clashing underfoot.

Polynesians settled New Zealand in 1250–1300 AD and developed a distinctive Māori culture, and Europeans first made contact in 1642 AD. The introduction of potatoes and muskets triggered upheaval among Māori early during the 19th century, which led to the inter-tribal Musket Wars. In 1840 the British and Māori signed a treaty making New Zealand a colony of the British Empire. Immigrant numbers increased sharply and conflicts escalated into the New Zealand Wars, which resulted in much Māori land being confiscated in the mid North Island. Economic depressions were followed by periods of political reform, with women gaining the vote during the 1890s, and a welfare state being established from the 1930s. After World War II, New Zealand joined Australia and the United States in the ANZUS security treaty, although the United States later suspended the treaty after New Zealand banned nuclear weapons. New Zealanders enjoyed one of the highest standards of living in the world in the 1950s, but the 1970s saw a deep recession, worsened by oil shocks and the United Kingdom's entry into the European Economic Community. The country underwent major economic changes during the 1980s, which transformed it from a protectionist to a liberalised free-trade economy. Markets for New Zealand's agricultural exports have diversified greatly since the 1970s, with once-dominant exports of wool being overtaken by dairy products, meat, and recently wine.

The majority of New Zealand's population is of European descent; the indigenous Māori are the largest minority, followed by Asians and non-Māori Polynesians. Māori and New Zealand Sign Language are the official languages, with English predominant. Much of New Zealand's culture is derived from Māori and early British settlers. Early European art was dominated by landscapes and to a lesser extent portraits of Māori. A recent resurgence of Māori culture has seen their traditional arts of carving, weaving and tattooing become more mainstream. Many artists now combine Māori and Western techniques to create unique art forms. The country's culture has also been broadened by globalisation and increased immigration from the Pacific Islands and Asia. New Zealand's diverse landscape provides many opportunities for outdoor pursuits and has provided the backdrop for a number of big budget movies.

New Zealand is organised into 11 regional councils and 67 territorial authorities for local government purposes; these have less autonomy than the country's long defunct provinces did. Nationally, executive political power is exercised by the Cabinet, led by the Prime Minister. Queen Elizabeth II is the country's head of state and is represented by a Governor-General. The Queen's Realm of New Zealand also includes Tokelau (a dependent territory); the Cook Islands and Niue (self-governing but in free association); and the Ross Dependency, New Zealand's territorial claim in Antarctica. New Zealand is a member of the Asia-Pacific Economic Cooperation, Commonwealth of Nations, Organisation for Economic Co-operation and Development, Pacific Islands Forum, and the United Nations.

Etymology

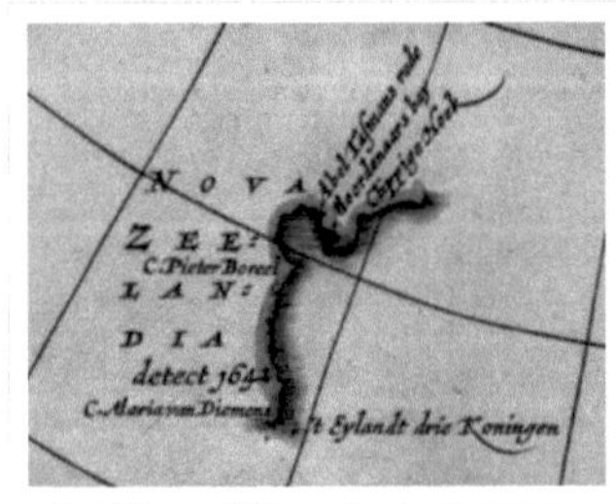
Detail from a 1657 map showing the western coastline of "Nova Zeelandia"

Aotearoa (often translated as "land of the long white cloud")[13] is the current Māori name for New Zealand, and is also used in New Zealand English. It is unknown whether Māori had a name for the whole country before the arrival of Europeans, with *Aotearoa* originally referring to just the North Island.[14] Abel Tasman sighted New Zealand in 1642 and called it *Staten Landt*, supposing it was connected to a landmass of the same name at the southern tip of South America.[15] In 1645 Dutch cartographers renamed the land *Nova Zeelandia* after the Dutch province of Zeeland.[16] [17] British explorer James Cook subsequently anglicised the name to New Zealand.[18]

Māori had several traditional names for the two main islands, including *Te Ika-a-Māui* (the fish of Māui) for the North Island and *Te Wai Pounamu* (the waters of greenstone) or *Te Waka o*

Aoraki (the canoe of Aoraki) for the South Island.[19] Early European maps labelled the islands North (North Island), Middle (South Island) and South (Stewart Island / Rakiura).[20] In 1830 maps began to use North and South to distinguish the two largest islands and by 1907 this was the accepted norm.[21] The New Zealand Geographic Board discovered in 2009 that the names of the North Island and South Island had never been formalised, but there are now plans to do so.[22] The board is also considering suitable Māori names,[23] with *Te Ika-a-Māui* and *Te Wai Pounamu* the most likely choices according to the chairman of the Māori Language Commission.[24]

History

New Zealand was one of the last major landmasses settled by humans. Radiocarbon dating, evidence of deforestation[26] and mitochondrial DNA variability within Māori populations[27] suggest New Zealand was first settled by Eastern Polynesians between 1250 and 1300,[19] [28] concluding a long series of voyages through the southern Pacific islands.[29] Over the centuries that followed these settlers developed a distinct culture now known as Māori. The population was divided into *iwi* (tribes) and *hapū* (subtribes) which would cooperate, compete and sometimes fight with each other. At some point a group of Māori migrated to the Chatham Islands (which they named *Rēkohu*) where they developed their distinct Moriori culture.[30] [31] The

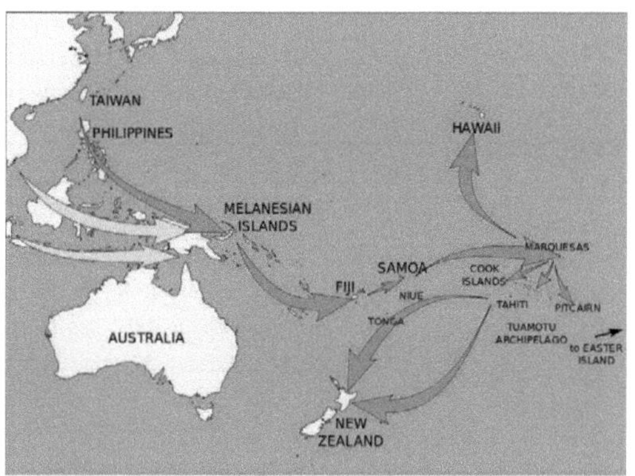

The Māori people are most likely descended from people that emigrated from Taiwan to Melanesia and then travelled east through to the Society Islands. After a pause of 70–265 years a new wave of exploration led to the discovery and settlement of New Zealand.[25]

Moriori population was decimated between 1835 and 1862, largely because of Māori invasion and enslavement, although European diseases also contributed. In 1862 only 101 survived and the last known full-blooded Moriori died in 1933.[32]

The first Europeans known to have reached New Zealand were Dutch explorer Abel Tasman and his crew in 1642.[33] In a hostile encounter, four crew members were killed and at least one Māori was hit by canister shot.[34] Europeans did not revisit New Zealand until 1769 when British explorer James Cook mapped almost the entire coastline.[33] Following Cook, New Zealand was visited by numerous European and North American whaling, sealing and trading ships. They traded food, metal tools, weapons and other goods for timber, food, artefacts, water, and on occasion sex.[35] The introduction of the potato and the musket transformed Māori agriculture and warfare. Potatoes provided a reliable food surplus, which enabled longer and more sustained military campaigns.[36] The resulting inter-tribal Musket Wars encompassed over 600 battles between 1801 and 1840, killing 30,000–40,000 Māori.[37] From the early 19th century, Christian missionaries began to settle New Zealand, eventually converting most of the Māori population.[38] The Māori population declined to around 40 percent of its pre-contact level during the 19th century; introduced diseases were the major factor.[39]

The Waitangi sheet from the Treaty of Waitangi

The British government appointed James Busby as British Resident to New Zealand in 1832[40] and in 1835, following an announcement of impending French sovereignty, the nebulous United Tribes of New Zealand sent a Declaration of the Independence to King William IV of the United Kingdom asking for protection.[40] Ongoing unrest and the dubious legal standing of the Declaration of Independence prompted the Colonial Office to send Captain William Hobson to claim sovereignty for the British Crown and negotiate a treaty with the Māori.[41] The Treaty of Waitangi was first signed in the Bay of Islands on 6 February 1840.[42] In response to the commercially run New Zealand Company's attempts to establish an independent settlement in Wellington[43] and French settlers "purchasing" land in Akaroa,[44] Hobson declared British sovereignty over all of New Zealand on 21 May 1840, even though copies of the Treaty were still circulating.[45] With the signing of the Treaty and declaration of sovereignty the number of immigrants, particularly from the United Kingdom, began to increase.[46]

New Zealand, originally part of the colony of New South Wales, became a separate Crown colony in 1841.[47] The colony gained a representative government in 1852 and the 1st New Zealand Parliament met in 1854.[48] In 1856 the colony effectively became self-governing, gaining responsibility over all domestic matters other than native policy. (Control over native policy was granted in the mid-1860s.)[48] Following concerns that the South Island might form a separate colony, premier Alfred Domett moved a resolution to transfer the capital from Auckland to a locality near the Cook Strait.[49] Wellington was chosen for its harbour and central location, with parliament officially sitting there for the first time in 1865. As immigrant numbers increased, conflicts over land led to the New Zealand Wars of the 1860s and 1870s, resulting in the loss and confiscation of much Māori land.[50] In 1893 the country became the first nation in the world to grant all women the right to vote[51] and in 1894 pioneered the adoption of compulsory arbitration between employers and unions.[52]

In 1907 New Zealand declared itself a Dominion within the British Empire and in 1947 the country adopted the Statute of Westminster, making New Zealand a Commonwealth realm.[48] New Zealand was involved in world affairs, fighting alongside the British Empire in the first and second World Wars[53] and suffering through the Great Depression.[54] The depression led to the election of the first Labour government and the establishment of a comprehensive welfare state and a protectionist economy.[55] New Zealand experienced increasing prosperity following World War II[56] and Māori began to leave their traditional rural life and move to the cities in search of work.[57] A Māori protest movement developed, which criticised Eurocentrism and worked for greater recognition of Māori culture and the Treaty of Waitangi.[58] In 1975, a Waitangi Tribunal was set up to investigate alleged breaches of the Treaty, and it was enabled to investigate historic grievances in 1985.[42] The government has negotiated settlements of these grievances with many iwi, although Māori claims to the foreshore and seabed have proved controversial in the 2000s.

Politics

Government

New Zealand is a constitutional monarchy with a parliamentary democracy,[59] although its constitution is not codified.[60] Queen Elizabeth II is the Queen of New Zealand and the head of state.[61] The Queen is represented by the Governor-General,[62] whom she appoints on the advice of the Prime Minister.[63] The Governor-General can exercise the Crown's prerogative powers (such as reviewing cases of injustice and making appointments of Cabinet ministers, ambassadors and other key public officials)[64] and in rare situations, the reserve powers (the power to dismiss a Prime Minister, dissolve Parliament or refuse the Royal Assent of a bill into law).[65] The powers of the Queen and the Governor-General are limited by constitutional constraints and they cannot normally be exercised without the advice of Cabinet.[65] [66]

John Key, the New Zealand Prime Minister

Elizabeth II

Sir Jerry Mateparae

The Parliament of New Zealand holds legislative power and consists of the Sovereign (represented by the Governor-General) and the House of Representatives.[66] It also included an upper house, the Legislative Council, until this was abolished in 1950.[66] The supremacy of the House over the Sovereign was established in England by the Bill of Rights 1689 and has been ratified as law in New Zealand.[66] The House of Representatives is democratically elected and a Government is formed from the party or coalition with the majority of seats.[66] If no majority is formed a minority government can be formed if support from other parties during confidence and supply votes is assured. The Governor-General appoints ministers under advice from the Prime Minister, who is by convention the Parliamentary leader of the governing party or coalition.[67] Cabinet, formed by ministers and led by the Prime Minister, is the highest policy-making body in government and responsible for deciding significant government actions.[68] By convention, members of cabinet are bound by collective responsibility to decisions made by cabinet.[69]

Judges and judicial officers are appointed non-politically and under strict rules regarding tenure to help maintain constitutional independence from the government.[60] This theoretically allows the judiciary to interpret the law based solely on the legislation enacted by Parliament without other influences on their decisions.[70] The Privy Council in London was the country's final court of appeal until 2004, when it was replaced with the newly established Supreme Court of New Zealand. The judiciary, headed by the Chief Justice,[71] includes the Court of Appeal, the High Court, and subordinate courts.[60]

Almost all parliamentary general elections between 1853 and 1996 were held under the first past the post voting system.[72] The elections since 1930 have been dominated by two political parties, National and Labour.[72] Since 1996, a form of proportional representation called Mixed Member Proportional (MMP) has been used.[60] Under the MMP system each person has two votes; one is for the 65 electoral seats (including seven reserved for Māori), and the other is for a party. The remaining 55 seats are assigned so that representation in parliament reflects the party vote, although a party has to win one electoral seat or 5 percent of the total party vote before it is eligible for these seats. Between March 2005 and August 2006 New Zealand became the only country in the world in which all the highest offices in the land (Head of State, Governor-General, Prime Minister, Speaker and Chief Justice) were occupied simultaneously by women.[73]

New Zealand government "Beehive" and the Parliament Buildings (right), in Wellington

Foreign relations and the military

Early colonial New Zealand allowed the British Government to determine external trade and be responsible for foreign policy.[74] The 1923 and 1926 Imperial Conferences decided that New Zealand should be allowed to negotiate their own political treaties, with the first successful commercial treaty being with Japan in 1928. Despite this independence New Zealand readily followed Britain in declaring war on Germany on 3 September 1939 with then Prime Minister Michael Savage proclaiming, "Where she goes, we go; where she stands, we stand."[75]

Māori Battalion haka in Egypt, 1941

In 1951 the United Kingdom became increasingly focused on its European interests,[76] while New Zealand joined Australia and the United States in the ANZUS security treaty.[77] The influence of the United States on New Zealand weakened following protests over the Vietnam War,[78] the failure of the United States to admonish France after the sinking of the Rainbow Warrior,[79] disagreements over environmental and agricultural trade issues and New Zealand's nuclear-free policy.[80] [81] Despite the USA's suspension of ANZUS obligations the treaty remained in effect between New Zealand and Australia, whose foreign policy has followed a similar historical trend.[82] Close political contact is maintained between the two countries, with free trade agreements and travel arrangements that allow citizens to visit, live and work in both countries without restrictions.[83] Currently over 500,000 New Zealanders live in Australia and 65,000 Australians live in New Zealand.[83]

New Zealand has a strong presence among the Pacific Island countries. A large proportion of New Zealand's aid goes to these countries and many Pacific people migrate to New Zealand for employment.[84] Permanent migration is regulated under the 1970 Samoan Quota Scheme and the 2002 Pacific Access Category, which allow up to 1,100 Samoan nationals and up to 750 other Pacific Islanders respectively to become permanent New Zealand residents each year. A seasonal workers scheme for temporary migration was introduced in 2007 and in 2009 about 8,000 Pacific Islanders were employed under it.[85] New Zealand is involved in the Pacific Islands Forum, Asia-Pacific Economic Cooperation and the Association of Southeast Asian Nations Regional Forum (including the East Asia Summit).[83] New Zealand is also a member of the United Nations,[86] the Commonwealth of Nations,[87] the Organisation for Economic Co-operation and Development[88] and the Five Powers Defence Arrangements.[89]

The New Zealand Defence Force has three branches: the Royal New Zealand Navy, the New Zealand Army and the Royal New Zealand Air Force.[90] New Zealand's national defence needs are modest because of the unlikelihood of direct attack,[91] although it does have a global presence. The country fought in both world wars, with notable campaigns in Gallipoli, Crete,[92] El Alamein[93] and Cassino.[94] The Gallipoli campaign played an important part in fostering New Zealand's national identity[95] [96] and strengthened the ANZAC tradition it shares with Australia.[97] According to Mary Edmond-Paul, "World War I had left scars on New Zealand society, with nearly 18,500 in total dying as a result of the war, more than 41,000 wounded,

Infantry from the 2nd Battalion, Auckland Regiment in the Battle of the Somme, September 1916.

and others affected emotionally, out of an overseas fighting force of about 103,000 and a population of just over a million."[98] New Zealand also played key parts in the naval Battle of the River Plate[99] and the Battle of Britain air campaign.[100] [101] During World War II, the United States had more than 400,000 American military personnel stationed in New Zealand.[102]

In addition to Vietnam and the two world wars, New Zealand fought in the Korean War, the Second Boer War,[103] the Malayan Emergency,[104] the Gulf War and the Afghanistan War. It has contributed forces to several regional and global peacekeeping missions, such as those in Cyprus, Somalia, Bosnia and Herzegovina, the Sinai, Angola, Cambodia, the Iran–Iraq border, Bougainville, East Timor, and the Solomon Islands.[105] New Zealand also sent a unit of army engineers to help rebuild Iraqi infrastructure for one year during the Iraq War.

Local government and external territories

The early European settlers divided New Zealand into provinces, which had a degree of autonomy.[106] Because of financial pressures and the desire to consolidate railways, education, land sales and other policies, government was centralised and the provinces were abolished in 1876.[107] As a result, New Zealand now has no separately represented subnational entities. The provinces are remembered in regional public holidays[108] and sporting rivalries.[109]

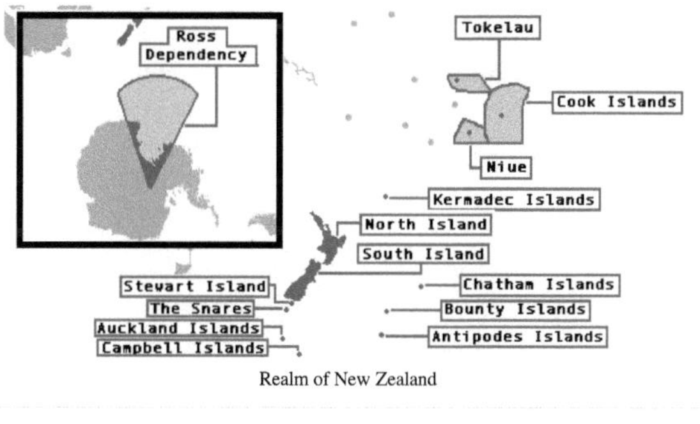

Realm of New Zealand

Since 1876, various councils have administered local areas under legislation determined by the central government.[106] [110] In 1989, the government reorganised local government into the current two-tier structure of regional councils and territorial authorities.[111] The 249 municipalities[111] that existed in 1975 have now been consolidated into 67 territorial authorities and 11 regional councils.[112] The regional councils' role is to regulate "the natural environment with particular emphasis on resource management",[111] while territorial authorities are responsible for sewage, water, local roads, building consents and other local matters.[113] Five of the territorial councils are unitary authorities and also act as regional councils.[114] The territorial authorities consist of 13 city councils, 53 district councils, and the Chatham Islands Council. While officially the Chatham Islands Council is not a unitary authority, it undertakes many functions of a regional council.[115]

The Realm of New Zealand is one of 16 realms within the commonwealth[116] [117] and comprises New Zealand, Tokelau, the Ross Dependency, the Cook Islands and Niue.[117] The Cook Islands and Niue are self-governing states in free association with New Zealand.[118] [119] The New Zealand Parliament cannot pass legislation for these countries, but with their consent can act on behalf of them in foreign affairs and defence. Tokelau is a non-self-governing territory that uses the New Zealand flag and anthem, but is administered by a council of three elders (one from each Tokelauan atoll).[120] [121] The Ross Dependency is New Zealand's territorial claim in Antarctica, where it operates the Scott Base research facility.[122] New Zealand citizenship law treats all parts of the realm equally, so most people born in New Zealand, the Cook Islands, Niue, Tokelau and the Ross Dependency before 2006 are New Zealand citizens. Further conditions apply for those born from 2006 onwards.[123]

Environment

Geography

    See also: Atlas of New Zealand at Wikimedia Commons

The snow-capped Southern Alps dominate the South Island, while the North Island's Northland Peninsula stretches towards the subtropics.

New Zealand is made up of two main islands and a number of smaller islands, located near the centre of the water hemisphere. The main North and South Islands are separated by the Cook Strait, 22 kilometres (14 mi) wide at its narrowest point.[124] Besides the North and South Islands, the five largest inhabited islands are Stewart Island, the Chatham Islands, Great Barrier Island (in the Hauraki Gulf),[125] d'Urville Island (in the Marlborough Sounds)[126] and Waiheke Island (about 22 km (14 mi) from central Auckland).[127] The country's islands lie between latitudes 29° and 53°S, and longitudes 165° and 176°E.

New Zealand is long (over 1600 kilometres (990 mi) along its north-north-east axis) and narrow (a maximum width of 400 kilometres (250 mi)),[128] with approximately 15134 km (9404 mi) of coastline[129] and a total land area of 268021 square kilometres (sq mi)[130] Because of its far-flung outlying islands and long coastline, the country has extensive marine resources. Its Exclusive Economic Zone, one of the largest in the world, covers more than 15 times its land area.[131]

The South Island is the largest land mass of New Zealand, and is divided along its length by the Southern Alps.[132] There are 18 peaks over 3000 metres (9800 ft), the highest of which is Aoraki/Mount Cook at 3754 metres (12316 ft).[133] Fiordland's steep mountains and deep fiords record the extensive ice age glaciation of this south-western corner of the South Island.[134] The North Island is less mountainous but is marked by volcanism.[135] The highly active Taupo volcanic zone has formed a large volcanic plateau, punctuated by the North Island's highest mountain, Mount Ruapehu (2797 metres (9177 ft)). The plateau also hosts the country's largest lake, Lake Taupo,[136] nestled in the caldera of one of the world's most active supervolcanoes.[137]

The country owes its varied topography, and perhaps even its emergence above the waves, to the dynamic boundary it straddles between the Pacific and Indo-Australian Plates.[138] New Zealand is part of Zealandia, a microcontinent nearly half the size of Australia that gradually submerged after breaking away from the Gondwanan supercontinent.[139] About 25 million years ago, a shift in plate tectonic movements began to contort and crumple the region. This is now most evident in the Southern Alps, formed by compression of the crust beside the Alpine Fault. Elsewhere the plate boundary involves the subduction of one plate under the other, producing the Puysegur Trench to the south, the Hikurangi Trench east of the North Island, and the Kermadec and Tonga Trenches[140] further north.[138]

Abel Tasman National Park in the South Island

Climate

New Zealand has a mild and temperate maritime climate with mean annual temperatures ranging from 10 °C (50 °F) in the south to 16 °C (61 °F) in the north.[141] Historical maxima and minima are 42.4 °C (108.3 °F) in Rangiora, Canterbury and −25.6 °C (−14.08 °F) in Ranfurly, Otago.[142] Conditions vary sharply across regions from extremely wet on the West Coast of the South Island to almost semi-arid in Central Otago and the Mackenzie Basin of inland Canterbury and subtropical in Northland.[143] Of the seven largest cities, Christchurch is the driest, receiving on average only 640 millimetres (25 in) of rain per year and Auckland the wettest, receiving almost twice that amount.[144] Auckland, Wellington and Christchurch all receive a yearly average in excess of 2,000 hours of sunshine. The southern and south-western parts of the South Island have a cooler and cloudier climate, with around 1,400–1,600 hours; the northern and north-eastern parts of the South Island are the sunniest areas of the country and receive approximately 2,400–2,500 hours.[145]

Biodiversity

The endemic flightless kiwi is a national icon.

New Zealand's geographic isolation for 80 million years[146] and island biogeography is responsible for the country's unique species of flora and fauna. They have either evolved from Gondwanan wildlife or the few organisms that have managed to reach the shores flying, swimming or being carried across the sea.[147] About 82 percent of New Zealand's indigenous vascular plants[148] are endemic, covering 1,944 species across 65 genera and includes a single family.[149] [150] The two main types of forest are those dominated by broadleaf trees with emergent podocarps, or by southern beech in cooler climates.[151] The remaining vegetation types consist of grasslands, the majority of which are tussock.[152]

Before the arrival of humans an estimated 80 percent of the land was covered in forest, with only high alpine, wet, infertile and volcanic areas without trees.[153] Massive deforestation occurred after humans arrived, with around half the forest cover lost to fire after Polynesian settlement.[154] Much of the remaining forest fell after European settlement, being logged or cleared to make room for pastoral farming, leaving forest occupying only 23 percent of the land.[155]

The forests were dominated by birds, and the lack of mammalian predators led to some like the kiwi, kakapo and takahē evolving flightlessness.[156] The arrival of humans, associated changes to habitat, and the introduction of rats, ferrets and other mammals led to the extinction of many bird species, including large birds like the moa and Haast's eagle.[157] [158]

Other indigenous animals are represented by reptiles (tuataras, skinks and geckos),[159] frogs, spiders (katipo), insects (weta) and snails.[160] [161] Some, such as the wrens and tuatara, are so unique that they have been called living fossils. Three species of bats (one since extinct) were the only sign of native land mammals in New Zealand until the 2006 discovery of bones from a unique, mouse-sized land mammal at least 16 million years old.[162] [163] Marine mammals however are abundant, with almost half the world's cetaceans (whales, dolphins, and porpoises) and large numbers of fur seals reported in New Zealand waters.[164] Many seabirds breed in New Zealand, a third of them unique to the country.[165] More penguin species are found in New Zealand than in any other country.[166]

Burnt forest near Levin, cleared for farming in 1909

Since human arrival almost half of the country's vertebrate species have become extinct, including at least fifty one birds, three frogs, three lizards, one freshwater fish, four plant species, and one bat.[157] Others are endangered or have had their range severely reduced.[157] However New Zealand conservationists have pioneered several methods to help threatened wildlife recover, including island sanctuaries, pest control, wildlife translocation, fostering, and ecological restoration of islands and other selected areas.[167] [168] [169] [170]

Economy

New Zealand has a modern, prosperous and developed market economy with an estimated gross domestic product (GDP) at purchasing power parity (PPP) per capita of roughly US$28,250.[171] The currency is the New Zealand dollar, informally known as the "Kiwi dollar"; it also circulates in the Cook Islands (see Cook Islands dollar), Niue, Tokelau, and the Pitcairn Islands.[172] New Zealand was ranked 5th in the 2011 Human Development Index,[173] 4th in the 2011 Index of Economic Freedom published by The Heritage Foundation.[174]

Milford Sound, one of New Zealand's most famous tourist destinations[175]

Historically, extractive industries have contributed strongly to New Zealand's economy, focussing at different times on sealing, whaling, flax, gold, kauri gum, and native timber.[176] With the development of refrigerated shipping in the 1880s meat and dairy products were exported to Britain, a trade which provided the basis for strong economic growth in New Zealand.[177] High demand for agricultural products from the United Kingdom and the United States helped New Zealanders achieve higher living standards than both Australia and Western Europe in the 1950s and 1960s.[178] In 1973 New Zealand's export market was reduced when the United Kingdom joined the European Community[179] and other compounding factors, such as the 1973 oil and 1979 energy crisis, led to a severe economic depression.[180] Living standards in New Zealand fell behind those of Australia and Western Europe, and by 1982 New Zealand had the lowest per-capita income of all the developed nations surveyed by the World Bank.[181] Since 1984, successive governments engaged in major macroeconomic restructuring (known first as Rogernomics and then Ruthanasia), rapidly transforming New Zealand from a highly protectionist economy to a liberalised free-trade economy.[182] [183]

Unemployment peaked above 10 percent in 1991 and 1992,[184] following the 1987 share market crash, but eventually fell a record low of 3.4 percent in 2007 (ranking fifth from twenty-seven comparable OECD nations).[185] The global financial crisis that followed however had a major impact on New Zealand with the GDP shrinking for five consecutive quarters, the longest recession in over thirty years,[186] [187] and unemployment rising back to 7% in late 2009.[188] The unemployment rate for youth was 17.4% in the June 2011 quarter.[189] New Zealand has experienced a series of "brain drains" since the 1970s[190] that still continue today.[191] Nearly one quarter of highly skilled workers live overseas, most in Australia and Britain, the most from any developed nation.[192] In recent years, however, a "brain gain" has brought in educated professionals from Europe and lesser developed countries.[193] [194]

Trade

New Zealand is heavily dependent on international trade,[195] particularly in agricultural products.[196] Exports account for a high 24 percent of its output,[129] making New Zealand vulnerable to international commodity prices and global economic slowdowns. Its principal export industries are agriculture, horticulture, fishing, forestry and mining, which make up about half of the country's exports.[197] Its major export partners are Australia, United States, Japan, China, and the United Kingdom.[129] On 7 April 2008, New Zealand and China signed the New Zealand China Free Trade Agreement, the first such agreement China has signed with a developed country.[198] [199] The service sector is the largest sector in the economy, followed by manufacturing and construction and then farming and raw material extraction.[129] Tourism plays a significant role in New Zealand's economy, contributing $15.0 billion to New Zealand's total GDP and supporting 9.6 percent of the total workforce in 2010.[200] International visitors to New Zealand increased by 3.1 percent in the year to October 2010[201] and are expected to increase at a rate of 2.5 percent annually up to 2015.[200]

Wool was New Zealand's major agricultural export during the late 19th century.[176] Even as late as the 1960s it made up over a third of all export revenues,[176] but since then its price has steadily dropped relative to other commodities[202] and wool is no longer profitable for many farmers.[203] In contrast dairy farming increased, with the number of dairy cows doubling between 1990 and 2007,[204] to become New Zealand's largest export earner.[205] In the year to June 2009, dairy products accounted for 21 percent ($9.1 billion) of total merchandise exports,[206] and the country's largest company, Fonterra, controls almost one-third of the international dairy trade.[207] Other agricultural exports in 2009 were meat 13.2 percent, wool 6.3 percent, fruit 3.5 percent and fishing 3.3 percent. New Zealand's wine industry has followed a similar trend to dairy, the number of vineyards doubling over the same period,[208] overtaking wool exports for the first time in 2007.[209] [210]

Wool has historically been one of New Zealand's major exports.

Infrastructure

In 2008, oil, gas and coal generated approximately 69 percent of New Zealand's gross energy supply and 31% was generated from renewable energy, primarily hydroelectric power and geothermal power.[211] New Zealand's transport network includes 93805 kilometres (58288 mi) of roads, worth 23 billion dollars,[212] and 4128 kilometres (2565 mi) of railway lines.[213] Most major cities and towns are linked by bus services, although the private car is the predominant mode of transport.[214] The railways were privatised in 1993, then re-purchased by the government in 2004 and vested into a state owned enterprise.[215] Railways run the length of the country, although most lines now carry freight rather than passengers.[216] Most international visitors arrive via air[217] and New Zealand has seven international airports, although currently only the Auckland and Christchurch airports connect directly with countries other than Australia or Fiji.[218] The New Zealand Post Office had a monopoly over telecommunications until 1989 when Telecom New Zealand was formed, initially as a state-owned enterprise and then privatised in 1990.[219] Telecom still owns the majority of the telecommunications infrastructure, but competition from other providers has increased.[220]

Demography

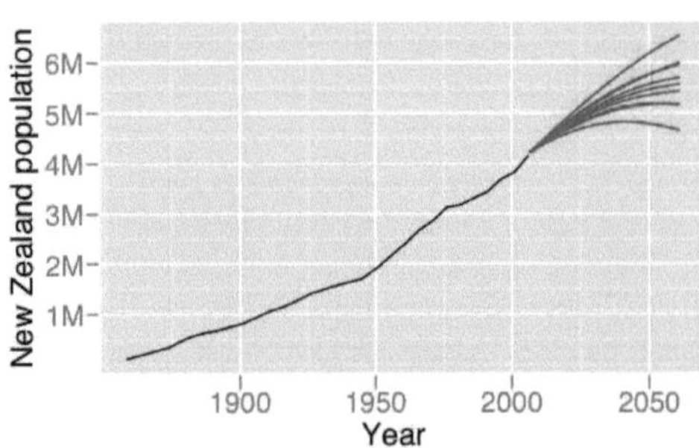

New Zealand's historical population (black) and projected growth (red)

The population of New Zealand is approximately 4.4 million.[221] New Zealand is a predominantly urban country, with 72 percent of the population living in 16 main urban areas and 53 percent living in the four largest cities of Auckland, Christchurch, Wellington, and Hamilton.[222] New Zealand cities generally rank highly on international livability measures. For instance, in 2010 Auckland was ranked the world's 4th most liveable city and Wellington the 12th by the Mercer Quality of Life Survey[223]

The life expectancy of a New Zealand child born in 2008 was 82.4 years for females, and 78.4 years for males.[224] Life expectancy at birth is forecast to increase from 80 years to 85 years in 2050 and infant mortality is expected to decline.[225] In 2050 the population is forecast to reach 5.3 million, the median age to rise from 36 years to 43 years and the percentage of people 60 years of age and older to rise from 18 percent to 29 percent.[225]

Ethnicity and immigration

In the 2006 census, 67.6 percent identified ethnically as European and 14.6 percent as Māori.[226] Other major ethnic groups include Asian (9.2 percent) and Pacific peoples (6.9 percent), while 11.1 percent identified themselves simply as a "New Zealander" (or similar) and 1 percent identified with other ethnicities.[227] [228] This contrasts with 1961, when the census reported that the population of New Zealand was 92 percent European and 7 percent Māori, with Asian and Pacific minorities sharing the remaining 1 percent.[229] While the demonym for a New Zealand citizen is New Zealander, the informal "Kiwi" is commonly used both internationally[230] and by locals.[231] The Māori loanword Pākehā usually refers to New Zealanders of European descent, although some reject this appellation,[232] [233] and some Māori use it to refer to all non-Polynesian New Zealanders.[234]

The Māori were the first people to reach New Zealand, followed by the early European settlers. Following colonisation, immigrants were predominantly from Britain, Ireland and Australia because of restrictive policies similar to the white Australian policies.[235] There was also significant Dutch, Dalmatian,[236] Italian, and German immigration together with indirect European immigration through Australia, North America, South America and South Africa.[237] Following the Great Depression policies were relaxed and migrant diversity increased. In 2009–10, an annual target of 45,000–50,000 permanent residence approvals was set by the New Zealand Immigration Service—more than one new migrant for every 100 New Zealand residents.[238] Twenty-three percent of New Zealand's population were born overseas, most of whom live in the Auckland region.[239] While most have still come from the United Kingdom and Ireland (29 percent), immigration from East Asia (mostly mainland China, but with substantial numbers also from Korea, Taiwan, Japan, and Hong Kong) is rapidly increasing the number of people from those countries.[240] The number of fee-paying international students increased sharply in the late 1990s, with more than 20,000 studying in public tertiary institutions in 2002.[241]

New Zealand's fastest growing ethnic groups are Asian. Here, lion dancers perform at the Auckland Lantern Festival.

Language

English is the predominant language in New Zealand, spoken by 98 percent of the population.[242] New Zealand English is similar to Australian English and many speakers from the Northern Hemisphere are unable to tell the accents apart.[243] After the Second World War, Māori were discouraged from speaking their own language (*te reo Māori*) in schools and workplaces and it existed as a community language only in a few remote areas.[244] It has recently undergone a process of revitalisation,[245] [246] being declared one of New Zealand's official languages in 1987,[247] and is spoken by 4.1 percent of the population.[242] There are now Māori language immersion schools and two Māori Television channels, the only nationwide television channels to have the majority of their prime-time content delivered in Māori.[248] Many places have officially been given dual Maori and English names in recent years. Samoan is one of the most widely spoken languages in New Zealand (2.3 percent),[249] followed by French, Hindi, Yue and Northern Chinese.[242] [250] [251] New Zealand Sign Language is used by approximately 28,000 people and was made New Zealand's second official language in 2006.[252]

A Ratana church

Education and religion

Primary and secondary schooling is compulsory for children aged 6 to 16, with the majority attending from the age of 5.[253] There are 13 school years and attending public schools is free. New Zealand has an adult literacy rate of 99 percent,[129] and over half of the population aged 15 to 29 hold a tertiary qualification.[253] [254] There are five types of government-owned tertiary institutions: universities, colleges of education, polytechnics, specialist colleges, and wānanga,[255] and also private training establishments.[256] In the adult population 14.2 percent have a bachelor's degree or higher, 30.4 percent have some form of secondary qualification as their highest qualification and 22.4 percent have no formal qualification.[257]

Christianity is the predominant religion in New Zealand. In the 2006 Census, 55.6 percent of the population identified themselves as Christians, while another 34.7 percent indicated that they had no religion (up from 29.6 percent in 2001) and around 4 percent affiliated with other religions.[258] [259] The main Christian denominations are Anglicanism, Roman Catholicism, Presbyterianism and Methodism. There are also significant numbers of Christians who identify themselves with Pentecostal, Baptist, and Latter-day Saint churches and the New Zealand-based Ratana church has adherents among Māori. According to census figures, other significant minority religions include Hinduism, Buddhism, and Islam.[250] [260]

Culture

Late twentieth-century house-post depicting the navigator Kupe fighting two sea creatures

Early Māori adapted the tropically based east Polynesian culture in line with the challenges associated with a larger and more diverse environment, eventually developing their own distinctive culture. Social organisation was largely communal with families (whanau), sub-tribes (hapu) and tribes (iwi) ruled by a chief (rangatira) whose position was subject to the community's approval.[262] The British and Irish immigrants brought aspects of their own culture to New Zealand and also influenced Māori culture,[263] [264] particularly with the introduction of Christianity.[265] However, Māori still regard their allegiance to tribal groups as a vital part of their identity, and Māori kinship roles resemble those of other Polynesian peoples.[266] More recently American, Australian, Asian and other European cultures have exerted influence on New Zealand. Non-Māori Polynesian cultures are also apparent, with Pasifika, the world's largest Polynesian festival, now an annual event in Auckland.

The largely rural life in early New Zealand led to the image of New Zealanders being rugged, industrious problem solvers.[267] Modesty was expected and enforced through the "tall poppy syndrome", where high achievers received harsh criticism.[268] At the time New Zealand was not known as an intellectual country.[269] From the early 20th century until the late 1960s Māori culture was suppressed by the attempted assimilation of Māori into British New Zealanders.[244] In the 1960s, as higher education became more available and cities expanded[270] urban culture began to dominate.[271] Even though the majority of the population now lives in cities, much of New Zealand's art, literature, film and humour has rural themes.

Cook Islands dancers at Auckland's Pasifika festival

Art

As part of the resurgence of Māori culture, the traditional crafts of carving and weaving are now more widely practised and Māori artists are increasing in number and influence.[272] Most Māori carvings feature human figures, generally with three fingers and either a natural-looking, detailed head or a grotesque head.[273] Surface patterns consisting of spirals, ridges, notches and fish scales decorate most carvings.[274] The pre-eminent Māori architecture consisted of carved meeting houses (wharenui) decorated with symbolic carvings and illustrations. These buildings were originally designed to be constantly rebuilt, changing and adapting to different whims or needs.[275]

Māori decorated the white wood of buildings, canoes and cenotaphs using red (a mixture of red ochre and shark fat) and black (made from soot) paint and painted pictures of birds, reptiles and other designs on cave walls.[276] Māori tattoos (moko) consisting of coloured soot mixed with gum were cut into the flesh with a bone chisel.[277] Since European arrival paintings and photographs have been dominated by landscapes, originally not as works of art but as factual portrayals of New Zealand.[278] Portraits of Māori were also common, with early painters often portraying them as "noble savages", exotic beauties or friendly natives.[278] The country's isolation delayed the influence of European artistic trends allowing local artists to developed their own distinctive style of regionalism.[279] During the 1960s and 70s many artists combined traditional Māori and Western techniques, creating unique art forms.[280] New Zealand art and craft has gradually achieved an international audience, with exhibitions in the Venice Biennale in 2001 and the "Paradise Now" exhibition in New York in 2004.[272] [281]

Māori cloaks are made of fine flax fibre and patterned with black, red and white triangles, diamonds and other geometric shapes.[282] Greenstone was fashioned into earrings and necklaces, with the most well-known design being the hei-tiki, a distorted human figure sitting cross-legged with its head tilted to the side.[283] Europeans brought English fashion etiquette to New Zealand, and until the 1950s most people dressed up for social occasions.[284] Standards have since relaxed and New Zealand fashion has received a reputation for being casual, practical and lacklustre.[285] [286] However, the local fashion industry has grown significantly since 2000, doubling exports and increasing from a handful to about 50 established labels, with some labels gaining international recognition.[286]

Portrait of Hinepare of Ngāti Kahungunu by Gottfried Lindauer, showing chin moko, pounamu hei-tiki and woven cloak

Literature

Māori quickly adopted writing as a means of sharing ideas, and many of their oral stories and poems were converted to the written form.[287] Most early English literature was obtained from Britain and it was not until the 1950s when local publishing outlets increased that New Zealand literature started to become widely known.[288] Although still largely influenced by global trends (modernism) and events (the Great Depression), writers in the 1930s began to develop stories increasingly focused on their experiences in New Zealand. During this period literature changed from a journalistic activity to a more academic pursuit.[289] Participation in the world wars gave some New Zealand writers a new perspective on New Zealand culture and with the post-war expansion of universities local literature flourished.[290]

Entertainment

New Zealand music has been influenced by blues, jazz, country, rock and roll and hip hop, with many of these genres given a unique New Zealand interpretation.[291] Māori developed traditional chants and songs from their ancient South-East Asian origins, and after centuries of isolation created a unique "monotonous" and "doleful" sound.[292] Flutes and trumpets were used as musical instruments[293] or as signalling devices during war or special occasions.[294] Early settlers brought over their ethnic music, with brass bands and choral music being popular, and musicians began touring New Zealand in the 1860s.[295] [296] Pipe bands became widespread during the early 20th century.[297] The New Zealand recording industry began to develop from 1940 onwards and many New Zealand musicians have obtained success in Britain and the USA.[291] Some artists release Māori language songs and the Māori tradition-based art of *kapa haka* (song and dance) has made a resurgence.[298]

Radio first arrived in New Zealand in 1922 and television in 1960, while the number of New Zealand films significantly increased during the 1970s.[299] In 1978 the New Zealand Film Commission started assisting local film-makers and many films attained a world audience, some receiving international acknowledgement. Deregulation in the 1980s saw a sudden increase in the numbers of radio and television stations.[299] New Zealand television primarily broadcasts American and British programming, along with a large number of Australian and local shows. The country's diverse scenery and compact size, plus government incentives,[300] have encouraged some producers to film big budget movies in New Zealand.[301] The New Zealand media industry is dominated by a small number of companies, most of which are foreign-owned, although the state retains ownership of some television and radio stations. Between 2003 and 2008, Reporters Without Borders consistently ranked New Zealand's press freedom in the top twenty.[302]

Sports

Statue of mountaineer Sir Edmund Hillary gazing towards Aoraki / Mount Cook

Most of the major sporting codes played in New Zealand have English origins.[303] Golf, netball, tennis and cricket are the four top participatory sports, soccer is the most popular among young people and rugby union attracts the most spectators.[304] Victorious rugby tours to Australia and the United Kingdom in the late 1880s and the early 1900s played an early role in instilling a national identity,[305] although the sport's influence has since declined.[306] Horse racing was also a popular spectator sport and became part of the "Rugby, Racing and Beer" culture during the 1960s.[307] Māori participation in European sports was particularly evident in rugby and the country's team performs a haka (traditional Māori challenge) before international matches.[308]

New Zealand has competitive international teams in rugby union, netball, cricket, rugby league, and softball and has traditionally done well in triathlons, rowing, yachting and cycling. The country has performed well on a medals-to-population ratio at Olympic Games and Commonwealth Games.[304] [309] New Zealand's national rugby union team is often regarded as the best in the world, and are the reigning World Cup holders. New Zealand are also the reigning rugby league world champions. New Zealand is known for its extreme sports, adventure tourism[310] and strong mountaineering tradition.[311] Other outdoor pursuits such as cycling, fishing, swimming, running, tramping, canoeing, hunting, snowsports and surfing are also popular.[312] The Polynesian sport of waka ama racing has increased in popularity and is now an international sport involving teams from all over the Pacific.[313]

See also

- International rankings of New Zealand

Notes

[1] "God Save the Queen" is officially a national anthem but is generally used only on regal and vice-regal occasions. "New Zealand's National Anthems" (http://www.mch.govt.nz/nz-identity-heritage/national-anthems). Ministry for Culture and Heritage. . Retrieved 17 February 2008.. Ministry for Culture and Heritage. . Retrieved 17 February 2008.

[2] Language percentages add to more than 100% because some people speak more than one language. They exclude unusable responses and those who spoke no language (e.g. too young to talk).

[3] Ethnicity percentages add to more than 100% because some people identify with more than one ethnic group.Didham, Robert; Potter, Deb (April 2005). *Understanding and Working with Ethnicity Data* (http://web.archive.org/web/20071125133402/http://www.stats.govt.nz/ NR/rdonlyres/F9967810-E15B-4D28-A8E3-DBAD6B80954C/0/UnderstandingWorkingEthnicityData.pdf). Statistics New Zealand. ISBN 9780478315059. Archived from the original (http://www.stats.govt.nz/browse_for_stats/population/census_counts/ review-measurement-of-ethnicity/~/media/Statistics/Publications/Analytical-reports/review-measurement-ethnicity/

understanding-working-ethnicity-data.ashx) on 25 November 2007. . Retrieved 19 September 2010.

[4] There is a multitude of dates that could be considered to mark independence (see Independence of New Zealand).

[5] The proportion of New Zealand's area (excluding estuaries) covered by rivers, lakes and ponds, based on figures from the New Zealand Land Cover Database, "The New Zealand Land Cover Database" (http://www.mfe.govt.nz/issues/land/land-cover-dbase/index.html). *New Zealand Land Cover Database 2*. New Zealand Ministry for the Environment. 1 July 2009. . Retrieved 26 April 2011. is (357526 + 81936) / (26821559 − 92499−26033 − 19216) = 1.6%. If estuarine open water, mangroves, and herbaceous saline vegetation are included, the figure is 2.2%.

[6] "National Population Estimates: September 2011 quarter" (http://www.stats.govt.nz/browse_for_stats/population/ estimates_and_projections/NationalPopulationEstimates_HOTPSep11qtr.aspx). Statistics New Zealand. 14 November 2011. . Retrieved 15 November 2011.

[7] "QuickStats About New Zealand's Population and Dwellings: Population counts" (http://stats.govt.nz/Census/2006CensusHomePage/ QuickStats/quickstats-about-a-subject/nzs-population-and-dwellings/population-counts.aspx). *2006 Census*. Statistics New Zealand. . Retrieved 14 April 2011.

[8] "New Zealand" (http://www.imf.org/external/pubs/ft/weo/2011/01/weodata/weorept.aspx?sy=2008&ey=2011&scsm=1&ssd=1& sort=country&ds=.&br=1&c=196&s=NGDPD,NGDPDPC,PPPGDP,PPPPC,LP&grp=0&a=&pr.x=28&pr.y=6). International Monetary Fund. . Retrieved 14 April 2011.

[9] "Equality and inequality: Gini index" (http://hdrstats.undp.org/en/indicators/161.html). *Human Development Report 2009*. United Nations Development Programme. . Retrieved 14 April 2011.

[10] "Human Development Report 2011" (http://hdr.undp.org/en/media/HDR_2011_EN_Table1.pdf). United Nations. . Retrieved 2011-11-02.

[11] The Chatham Islands have a separate time zone, 45 minutes ahead of the rest of New Zealand.

[12] The territories of Niue, the Cook Islands and Tokelau have their own cctlds, .nu, .ck and .tk respectively.

[13] King 2003, p. 41.

[14] Hay, Maclagan & Gordon 2008, p. 72.

[15] Wilson, John (March 2009). "European discovery of New Zealand − Tasman's achievement" (http://www.teara.govt.nz/en/ european-discovery-of-new-zealand/3). Te Ara - the Encyclopedia of New Zealand. . Retrieved 24 January 2011.

[16] Wilson, John (September 2007). "Tasman's achievement" (http://www.teara.govt.nz/en/european-discovery-of-new-zealand/3). Te Ara − the Encyclopedia of New Zealand. . Retrieved 16 February 2008.

[17] Mackay, Duncan (1986). "The Search For The Southern Land". In Fraser, B. *The New Zealand Book Of Events*. Auckland: Reed Methuen. pp. 52−54.

[18] Zeeland is spelt "Zealand" in English. New Zealand's name is not derived from the Danish island Zealand.

[19] Mein Smith 2005, p. 6.

[20] Brunner, Thomas (1851). *The Great Journey: an expedition to explore the interior of the Middle Island, New Zealand, 1846-8* (http://www. nzetc.org/tm/scholarly/BruJour-fig-BruJour_P001a.html). Royal Geographic Society. .

[21] McKinnon, Malcolm (November 2009). "Place names − Naming the country and the main islands" (http://www.TeAra.govt.nz/en/ place-names/1). Te Ara − the Encyclopedia of New Zealand. . Retrieved 24 January 2011.

[22] "Confusion over NZ islands' names" (http://news.bbc.co.uk/2/hi/asia-pacific/8011846.stm). *BBC News*. 22 April 2009. .

[23] May Eriksen, Alanah (25 April 2009). "Name quest unveils historic titles" (http://www.nzherald.co.nz/nz/news/article.cfm?c_id=1& objectid=10568595). *The New Zealand Herald*. .

[24] Davison, Isaac (22 April 2009). "North and South Islands officially nameless" (http://www.nzherald.co.nz/nz/news/article. cfm?c_id=1&objectid=10567873). *The New Zealand Herald*. .

[25] Wilmshurst, J. M.; Hunt, T. L.; Lipo, C. P.; Anderson, A. J. (2011). "High-precision radiocarbon dating shows recent and rapid initial human colonization of East Polynesia". *Proceedings of the National Academy of Sciences of the United States of America* **108** (5): 1815−1820. doi:10.1073/pnas.1015876108.

[26] McGlone, M. (1999). "Dating initial Maori environmental impact in New Zealand". *Quaternary International* **59**: 5−0. doi:10.1016/S1040-6182(98)00067-6.

[27] Murray-McIntosh, Rosalind P.; Scrimshaw, Brian J.; Hatfield, Peter J.; Penny, David (1998). "Testing migration patterns and estimating founding population size in Polynesia by using human mtDNA sequences". *Proceedings of the National Academy of Sciences of the United States of America* **95** (15): 9047−52. doi:10.1073/pnas.95.15.9047.

[28] Wilmshurst, J. M.; Anderson, A. J.; Higham, T. F. G.; Worthy, T. H. (2008). "Dating the late prehistoric dispersal of Polynesians to New Zealand using the commensal Pacific rat". *Proceedings of the National Academy of Sciences* **105**: 7676. Bibcode 2008PNAS..105.7676W. doi:10.1073/pnas.0801507105.

[29] Moodley, Y.; Linz, B.; Yamaoka, Y.; Windsor, H. M.; Breurec, S.; Wu, J. -Y.; Maady, A.; Bernhoft, S. et al. (2009). "The Peopling of the Pacific from a Bacterial Perspective". *Science* **323** (5913): 527. Bibcode 2009Sci...323..527M. doi:10.1126/science.1166083. PMC 2827536. PMID 19164753.

[30] Clark, Ross (1994). "Moriori and Māori: The Linguistic Evidence". In Sutton, Douglas. *The Origins of the First New Zealanders*. Auckland: Auckland University Press. pp. 123−135.

[31] Davis, Denise (September 2007). "The impact of new arrivals" (http://www.teara.govt.nz/en/moriori/4). Te Ara Encyclopedia of New Zealand. . Retrieved 30 April 2010.

[32] Davis, Denise; Solomon, Māui (March 2009). "'Moriori – The impact of new arrivals'" (http://www.TeAra.govt.nz/en/moriori/4). Te Ara – the Encyclopedia of New Zealand. . Retrieved 23 March 2011.

[33] Mein Smith 2005, p. 23.

[34] Salmond, Anne. *Two Worlds: First Meetings Between Maori and Europeans 1642–1772*. Auckland: Penguin Books. p. 82. ISBN 0670832987.

[35] King 2003, p. 122.

[36] Fitzpatrick, John (2004). "Food, warfare and the impact of Atlantic capitalism in Aotearo/New Zealand" (https://www.adelaide.edu.au/apsa/docs_papers/Others/Fitzpatrick.pdf). *Australasian Political Studies Association Conference: APSA 2004 Conference Papers*. .

[37] Brailsford, Barry (1972). *Arrows of Plague*. Wellington: Hick Smith and Sons. p. 35. ISBN 0456010602.

[38] Wagstrom, Thor (2005). "Broken Tongues and Foreign Hearts". In Brock, Peggy. *Indigenous Peoples and Religious Change*. Boston: Brill Academic Publishers. pp. 71 and 73. ISBN 9789004138995.

[39] Lange, Raeburn (1999). *May the people live: a history of Māori health development 1900–1920*. Auckland University Press. p. 18. ISBN 9781869402143.

[40] Rutherford, James (April 2009) [originally published in 1966]. "Busby, James" (http://www.teara.govt.nz/en/1966/busby-james/1). In McLintock, Alexander. *from An Encyclopaedia of New Zealand*. Te Ara – the Encyclopedia of New Zealand. . Retrieved 7 January 2011.

[41] McLintock, Alexander, ed (April 2009) [originally published in 1966]. "Sir George Gipps" (http://www.TeAra.govt.nz/en/1966/gipps-sir-george/1). *from An Encyclopaedia of New Zealand*. Te Ara – the Encyclopedia of New Zealand. . Retrieved 7 January 2011.

[42] Wilson, John (March 2009). "Government and nation – The origins of nationhood" (http://www.TeAra.govt.nz/en/government-and-nation/1). Te Ara – the Encyclopedia of New Zealand. . Retrieved 7 January 2011.

[43] McLintock, Alexander, ed (April 2009) [originally published in 1966]. "Settlement from 1840 to 1852" (http://www.teara.govt.nz/en/1966/land-settlement/3). *from An Encyclopaedia of New Zealand*. Te Ara – the Encyclopedia of New Zealand. . Retrieved 7 January 2011.

[44] Foster, Bernard (April 2009) [originally published in 1966]. "Akaroa, French Settlement At" (http://www.teara.govt.nz/en/1966/akaroa-french-settlement-at/1). In McLintock, Alexander. *from An Encyclopaedia of New Zealand*. Te Ara – the Encyclopedia of New Zealand. . Retrieved 7 January 2011.

[45] Simpson, K (September 2010). "Hobson, William – Biography" (http://www.teara.govt.nz/en/biographies/1h29/1). In McLintock, Alexander. *from the Dictionary of New Zealand Biography*. Te Ara – the Encyclopedia of New Zealand. . Retrieved 7 January 2011.

[46] Phillips, Jock (April 2010). "British immigration and the New Zealand Company" (http://www.teara.govt.nz/en/history-of-immigration/3). Te Ara – the Encyclopedia of New Zealand. . Retrieved 7 January 2011.

[47] "Crown colony era – the Governor-General" (http://www.nzhistory.net.nz/politics/history-of-the-governor-general/crown-colony-era). Ministry for Culture and Heritage. March 2009. . Retrieved 7 January 2011.

[48] Wilson, John (March 2009). "Government and nation – The constitution" (http://www.TeAra.govt.nz/en/government-and-nation/3). Te Ara – the Encyclopedia of New Zealand. . Retrieved 2 February 2011.

[49] Temple, Philip (1980). *Wellington Yesterday*. John McIndoe. ISBN 0-86868-012-5.

[50] "New Zealand's 19th-century wars – overview" (http://www.nzhistory.net.nz/war/new-zealands-19th-century-wars/introduction). Ministry for Culture and Heritage. April 2009. . Retrieved 7 January 2011.

[51] Wilson., John (March 2009). "History – Liberal to Labour" (http://www.TeAra.govt.nz/en/history/5). Te Ara – the Encyclopedia of New Zealand. . Retrieved 2 February 2011.

[52] Boxall, Peter; Haynes, Peter (1997). "Strategy and Trade Union Effectiveness in a Neo-liberal Environment" (http://www.gurn.info/en/topics/global-trade-union-strategies-union-renewal/organizational-innovation-and-change/industrial-relations-and-labour-regulations-affecting-unions2019-structure/strategy-and-trade-union-effectiveness-in-a-neo-liberal-environment) (PDF). *British Journal of Industrial Relations* **35** (4): 567–591. doi:10.1111/1467-8543.00069. .

[53] "War and Society" (http://www.nzhistory.net.nz/war-and-society). Ministry for Culture and Heritage. . Retrieved 7 January 2011.

[54] Easton, Brian (April 2010). "Economic history – Interwar years and the great depression" (http://www.TeAra.govt.nz/en/economic-history/7). Te Ara – the Encyclopedia of New Zealand. . Retrieved 7 January 2011.

[55] Derby, Mark (May 2010). "Strikes and labour disputes – Wars, depression and first Labour government" (http://www.TeAra.govt.nz/en/strikes-and-labour-disputes/6). Te Ara – the Encyclopedia of New Zealand. . Retrieved 1 February 2011.

[56] Easton, Brian (November 2010). "Economic history – Great boom, 1935–1966" (http://www.TeAra.govt.nz/en/economic-history/9). Te Ara – the Encyclopedia of New Zealand. . Retrieved 1 February 2011.

[57] Keane, Basil (November 2010). "Te Māori i te ohanga – Māori in the economy – Urbanisation" (http://www.TeAra.govt.nz/en/te-maori-i-te-ohanga-maori-in-the-economy/6). Te Ara – the Encyclopedia of New Zealand. . Retrieved 7 January 2011.

[58] Royal, Te Ahukaramū (March 2009). "Māori – Urbanisation and renaissance" (http://www.teara.govt.nz/en/maori/5). Te Ara – the Encyclopedia of New Zealand. . Retrieved 1 February 2011.

[59] "Queen and New Zealand" (http://www.royal.gov.uk/MonarchAndCommonwealth/NewZealand/NewZealand.aspx). The British Monarchy. . Retrieved 28 April 2010.

[60] "Factsheet – New Zealand – Political Forces" (http://web.archive.org/web/20060514204533/http://economist.com/countries/NewZealand/profile.cfm?folder=Profile-Political Forces). *The Economist*. 15 February 2005. . Retrieved 4 August 2009.

[61] "New Zealand Legislation: Royal Titles Act 1974" (http://www.legislation.govt.nz/act/public/1974/0001/latest/DLM411814.html). New Zealand Government. February 1974. . Retrieved 8 January 2011.

[62] "The Governor General of New Zealand" (http://www.gg.govt.nz/). Official website of the Governor General. . Retrieved 8 January 2011.

[63] "The Queen's role in New Zealand" (http://www.royal.gov.uk/MonarchAndCommonwealth/NewZealand/TheQueensroleinNewZealand.aspx). The British Monarchy. . Retrieved 28 April 2010.

[64] Harris, Bruce (2009). "Replacement of the Royal Prerogative in New Zealand" (http://www.britannica.com/bps/additionalcontent/18/41876855/REPLACEMENT-OF-THE-ROYAL-PREROGATIVE-IN-NEW-ZEALAND). *New Zealand Universities Law Review* **23**: 285–314. .

[65] "The Reserve Powers" (http://www.gg.govt.nz/role/powers.htm). Governor General. . Retrieved 8 January 2011.

[66] "How Parliament works: What is Parliament?" (http://www.parliament.nz/en-NZ/AboutParl/HowPWorks/FactSheets/0/e/7.htm). New Zealand Parliament. 28 June 2010. . Retrieved 8 January 2011.

[67] "How Parliament works: People in Parliament" (http://www.parliament.nz/en-NZ/AboutParl/HowPWorks/People/0/a/f/0afe831bd86b4c03b92cc769806bdb3b.htm#_Toc139097923). New Zealand Parliament. August 2006. . Retrieved 9 January 2011.

[68] Wilson, John (November 2010). "Government and nation – System of government" (http://www.TeAra.govt.nz/en/government-and-nation/4). Te Ara – the Encyclopedia of New Zealand. . Retrieved 9 January 2011.

[69] "Cabinet Manual: Cabinet" (http://www.cabinetmanual.cabinetoffice.govt.nz/5.2). Department of Prime Minister and Cabinet. 2008. . Retrieved 2 March 2011.

[70] "The Judiciary" (http://www.justice.govt.nz/courts/the-judiciary). Ministry of Justice. . Retrieved 9 January 2011.

[71] "The Current Chief Justice" (http://www.courtsofnz.govt.nz/about/judges/current-chief). Courts of New Zealand. . Retrieved 9 January 2011.

[72] "First past the post – the road to MMP" (http://www.nzhistory.net.nz/politics/fpp-to-mmp/first-past-the-post). Ministry for Culture and Heritage. September 2009. . Retrieved 9 January 2011.

[73] Collins, Simon (May 2005). "Women run the country but it doesn't show in pay packets" (http://www.nzherald.co.nz/nz/news/article.cfm?c_id=1&objectid=10127960). *The New Zealand Herald*. .

[74] McLintock, Alexander, ed (April 2009) [originally published in 1966]. "External Relations" (http://www.teara.govt.nz/en/1966/history-constitutional/10). *from An Encyclopaedia of New Zealand*. Te Ara – the Encyclopedia of New Zealand. . Retrieved 7 January 2011.

[75] "Michael Joseph Savage" (http://www.nzhistory.net.nz/people/michael-joseph-savage-biography). Ministry for Culture and Heritage. July 2010. . Retrieved 29 January 2011.

[76] Patman, Robert (2005). "Globalisation, Sovereignty, and the Transformation of New Zealand Foreign Policy" (http://www.victoria.ac.nz/css/docs/Working_Papers/WP21.pdf) (PDF). *Working Paper 21/05*. Centre for Strategic Studies, Victoria University of Wellington. p. 8. . Retrieved 12 March 2007.

[77] "Department Of External Affairs: Security Treaty between Australia, New Zealand and the United States of America" (http://www.australianpolitics.com/foreign/anzus/anzus-treaty.shtml). Australian Government. September 1951. . Retrieved 11 January 2011.

[78] "The Vietnam War" (http://www.nzhistory.net.nz/war/vietnam-war). Ministry for Culture and Heritage. June 2008. . Retrieved 11 January 2011.

[79] "Sinking the Rainbow Warrior – nuclear-free New Zealand" (http://www.nzhistory.net.nz/politics/nuclear-free-new-zealand/rainbow-warrior). Ministry for Culture and Heritage. August 2008. . Retrieved 11 January 2011.

[80] "Nuclear-free legislation – nuclear-free New Zealand" (http://www.nzhistory.net.nz/politics/nuclear-free-new-zealand/nuclear-free-zone). New Zealand History Online. August 2008. . Retrieved 11 January 2011.

[81] Lange, David (1990). *Nuclear Free: The New Zealand Way*. New Zealand: Penguin Books. ISBN 0140145192.

[82] "Australia in brief" (http://www.dfat.gov.au/aib/history.html). Australian Department of Foreign Affairs and Trade. . Retrieved 11 January 2011.

[83] "New Zealand country brief" (http://www.dfat.gov.au/geo/new_zealand/nz_country_brief.html). Department of Foreign Affairs and Trade. . Retrieved 11 January 2011.

[84] Bertram, Geoff (April 2010). "South Pacific economic relations – Aid, remittances and tourism" (http://www.TeAra.govt.nz/en/south-pacific-economic-relations/4). Te Ara – the Encyclopedia of New Zealand. . Retrieved 11 January 2011.

[85] Howes, Stephen (November 2010). "Making migration work: Lessons from New Zealand" (http://devpolicy.org/making-migration-work-lessons-from-new-zealand/). Development Policy Centre. . Retrieved 23 March 2011.

[86] "Member States of the United Nations" (http://www.un.org/en/members/index.shtml#n). United Nations. . Retrieved 11 January 2011.

[87] "The Commonwealth in the Pacific" (http://www.commonwealth-of-nations.org/Home-Pacific,56,44,1). Commonwealth of Nations. . Retrieved 11 January 2011.

[88] "Members and partners" (http://www.oecd.org/pages/0,3417,en_36734052_36761800_1_1_1_1_1,00.html). Organisation for Economic Co-operation and Development. . Retrieved 11 January 2011.

[89] "New Zealand Embassy Washington, United States of America: Defence relations" (http://www.nzembassy.com/usa/relationship-between-new-zealand-and-usa/new-zealand-and-usa/defence-relations). New Zealand Ministry of Foreign Affairs and Trade. . Retrieved 11 January 2011.

[90] "Welcome to NZDF" (http://www.nzdf.mil.nz/default.htm). New Zealand Defence Force. . Retrieved 11 January 2011.

[91] Ayson, Robert (2007). "New Zealand Defence and Security Policy,1990–2005". In Alley, Roderic. *New Zealand In World Affairs, Volume IV: 1990–2005*. Wellington: Victoria University Press. p. 132. ISBN 9780864735485.

[92] "The Battle for Crete" (http://www.nzhistory.net.nz/war/the-battle-for-crete). Ministry for Culture and Heritage. May 2010. . Retrieved 9 January 2011.

[93] "El Alamein – The North African Campaign" (http://www.nzhistory.net.nz/war/the-north-african-campaign/el-alamein). Ministry for Culture and Heritage. May-2009. . Retrieved 9 January 2011.

[94] Holmes, Richard (September 2010). "World War Two: The Battle of Monte Cassino" (http://www.bbc.co.uk/history/worldwars/wwtwo/battle_cassino_01.shtml). . Retrieved 9 January 2011.

[95] "Gallipoli stirred new sense of national identity says Clark" (http://www.nzherald.co.nz/nz/news/article.cfm?c_id=1&objectid=10122323). *New Zealand Herald*. April 2005. .

[96] Prideaux, Bruce (2007). Ryan, Chris. ed. *Battlefield tourism: history, place and interpretation*. Elsevier Science. p. 18. ISBN 978-0080453620.

[97] Burke, Arthur. "The Spirit of ANZAC" (http://www.anzacday.org.au/spirit/spirit2.html). ANZAC Day Commemoration Committee. . Retrieved 11 January 2011.

[98] Mary Edmond-Paul (2008). *Lighted windows: critical essays on Robin Hyde* . Otago University Press. p.77. ISBN 1877372587

[99] "New Zealand and the Battle of River Plate" (http://www.mfat.govt.nz/Foreign-Relations/Latin-America/News/0-river-plate.php). New Zealand Ministry of Foreign Affairs and Trade. . Retrieved 29 January 2011.

[100] "Airmen from New Zealand who took part in the Battle of Britain" (http://www.bbm.org.uk/pilots-nz.htm). The Battle of Britain London Monument. . Retrieved 10 January 2011.

[101] "New Zealand's contribution – The Battle of Britain" (http://www.nzhistory.net.nz/war/battle-of-britain/kiwi-contribution). Ministry for Culture and Heritage. September 2010. . Retrieved 10 January 2011.

[102] "Bureau of East Asian and Pacific Affairs Background Note: New Zealand" (http://www.state.gov/r/pa/ei/bgn/35852.htm). US Department of State. August 2010. . Retrieved 10 January 2011.

[103] "South African War 1899–1902" (http://www.nzhistory.net.nz/war/the-south-african-boer-war/introduction). Ministry for Culture and Heritage. February 2009. . Retrieved 11 January 2011.

[104] "NZ and the Malayan Emergency" (http://www.nzhistory.net.nz/war/the-malayan-emergency). Ministry for Culture and Heritage. August 2010. . Retrieved 11 January 2011.

[105] "New Zealand Defence Force Overseas Operations" (http://web.archive.org/web/20080125104529/http://www.nzdf.mil.nz/operations/default.htm). New Zealand Defence Force. January 2008. Archived from the original (http://www.nzdf.mil.nz/operations/default.htm) on 25 January 2008. . Retrieved 17 February 2008.

[106] "New Zealand's Nine Provinces (1853–76)" (http://www.library.otago.ac.nz/pdf/hoc_fr_bulletins/31_bulletin.pdf). Friends of the Hocken Collections. March 2000. . Retrieved 13 January 2011.

[107] McLintock, Alexander, ed (April 2009) [originally published in 1966]. "Provincial Divergencies" (http://www.TeAra.govt.nz/en/1966/provinces-and-provincial-districts/3). *from An Encyclopaedia of New Zealand*. Te Ara – the Encyclopedia of New Zealand. . Retrieved 7 January 2011.

[108] "Public holidays" (http://www.dol.govt.nz/er/holidaysandleave/publicholidays/index.asp). New Zealand Department of Labour. . Retrieved 2 April 2011.

[109] "Overview – regional rugby" (http://www.nzhistory.net.nz/culture/regional-rugby/overview). Ministry for Culture and Heritage. September 2010. . Retrieved 13 January 2011.

[110] Dollery, Brian; Keogh, Ciaran; Crase, Lin (2007). "Alternatives to Amalgamation in Australian Local Government: Lessons from the New Zealand Experience" (http://www.anzrsai.org/system/files/f8/f9/f39/f40/o186//Dollery sustaining regions article.pdf). *Sustaining Regions* **6** (1): 50–69. .

[111] Sancton, Andrew (2000). *Merger mania: the assault on local government*. McGill-Queen's University Press. p. 84. ISBN 0773521631.

[112] "Subnational population estimates at 30 June 2010 (boundaries at 1 November 2010)" (http://web.archive.org/web/20110610051916/http://www.stats.govt.nz/~/media/Statistics/Methods and Services/Tables/Subnational population estimates/subpopest2001-10.ashx). Statistics New Zealand. 26 October 2010. . Retrieved 2 April 2011.

[113] Smelt, Roselynn; Jui Lin, Yong (2009). *New Zealand*. Cultures of the World (2nd ed.). New York: Marshall Cavendish. p. 33. ISBN 9780761434153.

[114] "Unitary Authority" (http://www.fndc.govt.nz/your-council/unitary-authority). Far North District Council. . Retrieved 29 January 2011.

[115] "Minutes of the Statutory Meeting of the Chatham Islands Council" (http://www.cic.govt.nz/pdfs/councilMeetings/cic-council-statutory-minutes-281010.pdf). Chatham Islands Council. October 2010. . Retrieved 29 January 2011.

[116] "What is a Commonwealth Realm?" (http://www.royal.gov.uk/MonarchAndCommonwealth/QueenandCommonwealth/WhatisaCommonwealthRealm.aspx). Royal Household. . Retrieved 6 October 2009.

[117] "New Zealand's Constitution" (http://www.gg.govt.nz/role/constofnz.htm). The Governor-General of New Zealand. . Retrieved 13 January 2010.

[118] "System of Government" (http://www.gov.nu/wb/pages/system-of-government-fakatokaaga-he-fakatufono.php). Government of Niue. . Retrieved 13 January 2010.

[119] "Government – Structure, Personnel" (http://www.ck/govt.htm#con). Government of the Cook Islands. . Retrieved 13 January 1010.

[120] "Tourism, Travel, & Information Guide to the New Zealand Territory of Tokelau" (http://www.tokelau.com/). Tokelau.com. . Retrieved 13 January 2010.

[121] "Government" (http://www.tokelau.org.nz/Tokelau+Government/Government.html). Tokelau Government. . Retrieved 13 January 2010.

[122] "Scott Base" (http://www.antarcticanz.govt.nz/scott-base). Antarctica New Zealand. . Retrieved 13 January 2010.

[123] "Am I a New Zealand Citizen?" (http://www.dia.govt.nz/Services-Citizenship-Am-I-a-New-Zealand-Citizen?OpenDocument). New Zealand Department of Internal Affairs. . Retrieved 3 March 2011.

[124] McLintock, Alexander, ed (April 2009) [originally published in 1966]. "The Sea Floor" (http://www.TeAra.govt.nz/en/1966/cook-strait/1). *from An Encyclopaedia of New Zealand*. Te Ara – the Encyclopedia of New Zealand. . Retrieved 13 January 2011.

[125] "Hauraki Gulf islands" (http://www.aucklandcity.govt.nz/auckland/introduction/hauraki/default.asp). Auckland City Council. . Retrieved 13 January 2011.

[126] Hindmarsh (2006). "Discovering D'Urville" (http://www.historic.org.nz/en/Publications/HeritageNZMagazine/HeritageNz2006/HNZ06-DiscoveringDUrville.aspx). Heritage New Zealand. . Retrieved 13 January 2011.

[127] "Distance tables" (http://www.auckland-coastguard.org.nz/Information/Distance+Tables.html). Auckland Coastguard. . Retrieved 2 March 2011.

[128] McKenzie, D. W. (1987). *Heinemann New Zealand atlas*. Heinemann Publishers. ISBN 079000187X.

[129] "The World Factbook – New Zealand" (https://www.cia.gov/library/publications/the-world-factbook/geos/nz.html). *CIA*. 15 November 2007. . Retrieved 30 November 2007.

[130] "Geography" (http://www2.stats.govt.nz/domino/external/PASFull/pasfull.nsf/84bf91b1a7b5d7204c256809000460a4/4c2567ef00247c6acc25697a00043f15?OpenDocument). Statistics New Zealand. 1999. . Retrieved 21 December 2009.

[131] (PDF) *Offshore Options: Managing Environmental Effects in New Zealand's Exclusive Economic Zone* (http://www.mfe.govt.nz/publications/oceans/offshore-options-jun05/offshore-options-jun05.pdf). Wellington: Ministry for the Environment. 2005. ISBN 0-478-25916-6. .

[132] Coates, Glen (2002). *The rise and fall of the Southern Alps*. Canterbury University Press. p. 15. ISBN 0908812930.

[133] Garden 2005, p. 52.

[134] Grant, David (March 2009). "Southland places – Fiordland's coast" (http://www.TeAra.govt.nz/en/southland-places/10). Te Ara – the Encyclopedia of New Zealand. . Retrieved 14 January 2011.

[135] "Central North Island volcanoes" (http://www.doc.govt.nz/parks-and-recreation/national-parks/tongariro/features/central-north-island-volcanoes/). Department of Conservation. . Retrieved 14 January 2011.

[136] Walrond, Carl (March 2009). "Natural environment – Geography and geology" (http://www.teara.govt.nz/en/natural-environment/1). Te Ara – the Encyclopedia of New Zealand. . Retrieved 14 January 2010.

[137] "Taupo" (http://www.geonet.org.nz/volcano/activity/taupo/about.html). GNS Science. . Retrieved 2 April 2011.

[138] Lewis, Keith; Nodder, Scott; Carter, Lionel (March 2009). "Sea floor geology – Active plate boundaries" (http://www.TeAra.govt.nz/en/sea-floor-geology/2). Te Ara – the Encyclopedia of New Zealand. . Retrieved 4 February 2011.

[139] Wallis, G. P.; Trewick, S. A. (2009). "New Zealand phylogeography: evolution on a small continent". *Molecular Ecology* **18** (17): 3548–3580. doi:10.1111/j.1365-294X.2009.04294.x. PMID 19674312.

[140] Wright, Dawn; Bloomer, Sherman; MacLeod, Christopher; Taylor, Brian; Goodliffe, Andrew (2000). "Bathymetry of the Tonga Trench and Forearc: A Map Series". *Marine Geophysical Researches* **21** (5): 489–512. doi:10.1023/A:1026514914220.

[141] Mullan, Brett; Tait, Andrew; Thompson, Craig (March 2009). "Climate – New Zealand's climate" (http://www.TeAra.govt.nz/en/climate/1). Te Ara – the Encyclopedia of New Zealand. . Retrieved 15 January 2011.

[142] "Summary of New Zealand climate extremes" (http://www.niwa.co.nz/education-and-training/schools/resources/climate/extreme). National Institute of Water and Atmospheric Research. 2004. . Retrieved 30 April 2010.

[143] Walrond, Carl (March 2009). "Natural environment – Climate" (http://www.TeAra.govt.nz/en/natural-environment/3). Te Ara – the Encyclopedia of New Zealand. . Retrieved 15 January 2011.

[144] "Mean monthly rainfall" (http://web.archive.org/web/20110503221956/http://www.niwa.co.nz/__data/assets/file/0006/44268/rain.xls) (XLS). National Institute of Water and Atmospheric Research. . Retrieved 4 February 2011.

[145] "Mean monthly sunshine hours" (http://web.archive.org/web/20110610201128/http://www.niwascience.co.nz/__data/assets/file/0006/44655/sunshine.xls) (XLS). National Institute of Water and Atmospheric Research. . Retrieved 4 February 2011.

[146] Cooper, R.; Millener, P. (1993). "The New Zealand biota: Historical background and new research". *Trends in Ecology & Evolution* **8**: 429. doi:10.1016/0169-5347(93)90004-9.

[147] Lindsey, Terence; Morris, Rod (2000). *Collins Field Guide to New Zealand Wildlife*. HarperCollins (New Zealand) Limited. p. 14. ISBN 9781869503000.

[148] New Zealand has approximately 4,000 indigenous species of lichens and other non-vascular plants, and only 40 percent of these are endemic.Wassilieff, Maggy (March 2009). "Lichens – Lichens in New Zealand" (http://www.TeAra.govt.nz/en/lichens/2). Te Ara – the Encyclopedia of New Zealand. . Retrieved 16 January 2011.

[149] "Frequently asked questions about New Zealand plants" (http://www.nzpcn.org.nz/page.asp?help_faqs_NZ_plants). New Zealand Plant Conservation Network. May 2010. . Retrieved 15 January 2011.

[150] Rolfe, Peter; Sawyer, John (2006). *New Zealand indigenous vascular plant checklist*. New Zealand Plant Conservation Network. ISBN 0-473-11306-6.

[151] McLintock, Alexander, ed (April 2010) [originally published in 1966]. "Mixed Broadleaf Podocarp and Kauri Forest" (http://www.TeAra.govt.nz/en/1966/forests-indigenous/4). *from An Encyclopaedia of New Zealand*. Te Ara – the Encyclopedia of New Zealand. .

Retrieved 15 January 2011.

[152] Mark, Alan (March 2009). "Grasslands − Tussock grasslands" (http://www.TeAra.govt.nz/en/grasslands/1). Te Ara − the Encyclopedia of New Zealand. . Retrieved 17 January 2010.

[153] "Commentary on Forest Policy in the Asia-Pacific Region (A Review for Indonesia, Malaysia, New Zealand, Papua New Guinea, Philippines, Thailand and Western Samoa)" (http://www.fao.org/docrep/w7730e/w7730e09.htm#new zealand). Forestry Department. 1997. . Retrieved 4 February 2011.

[154] McGlone, M.S. (1989). "The Polynesian settlement of New Zealand in relation to environmental and biotic changes" (http://nzes.org.nz/nzje/free_issues/NZJEcol12_s_115.pdf). *New Zealand Journal of Ecology* **12(S)**: 115−129. .

[155] Taylor, R. and Smith, I. (1997). The state of New Zealand's environment 1997 (http://www.mfe.govt.nz/publications/ser/ser1997/index.html). Ministry for the Environment, Wellington.

[156] "New Zealand ecology: Flightless birds" (http://www.terranature.org/flightlessbirds.htm). TerraNature. . Retrieved 17 January 2011.

[157] Holdaway, Richard (March 2009). "Extinctions − New Zealand extinctions since human arrival" (http://www.TeAra.govt.nz/en/extinctions/4). Te Ara − the Encyclopedia of New Zealand. . Retrieved 4 February 2011.

[158] Kirby, Alex (January 2005). "Huge eagles 'dominated NZ skies'" (http://news.bbc.co.uk/2/hi/science/nature/4138147.stm). BBC News. .

[159] "Tuatara: New Zealand reptiles" (http://www.doc.govt.nz/conservation/native-animals/reptiles-and-frogs/tuatara/). Department of Conservation. . Retrieved 17 January 2011.

[160] Ryan, Paddy (March 2009). "Snails and slugs − Flax snails, giant snails and veined slugs" (http://www.TeAra.govt.nz/en/snails-and-slugs/2). Te Ara − the Encyclopedia of New Zealand. . Retrieved 4 February 2011.

[161] "Native Animals" (http://www.doc.govt.nz/conservation/native-animals/). Department of Conservation. . Retrieved 17 January 2011.

[162] "Tiny Bones Rewrite Textbooks, first New Zealand land mammal fossil" (http://web.archive.org/web/20070531085218/http://www.science.unsw.edu.au/news/2006/nzmammal.html). University of New South Wales. 31 May 2007. Archived from the original (http://www.science.unsw.edu.au/news/2006/nzmammal.html) on 31 May 2007. .

[163] Worthy, Trevor H.; Tennyson, Alan J. D.; Archer, Michael; Musser, Anne M.; Hand, Suzanne J.; Jones, Craig; Douglas, Barry J.; McNamara, James A. et al. (2006). "Miocene mammal reveals a Mesozoic ghost lineage on insular New Zealand, southwest Pacific". *Proceedings of the National Academy of Sciences of the United States of America* **103** (51): 19419−23. doi:10.1073/pnas.0605684103.

[164] "Marine Mammals" (http://www.doc.govt.nz/conservation/native-animals/marine-mammals/). Department of Conservation. . Retrieved 17 January 2011.

[165] "Sea & shore birds" (http://www.doc.govt.nz/conservation/native-animals/birds/sea-and-shore-birds/). New Zealand Department of Conservation. . Retrieved 7 March 2011.

[166] "Penguins" (http://www.doc.govt.nz/conservation/native-animals/birds/sea-and-shore-birds/penguins/penguins/). New Zealand Department of Conservation. . Retrieved 7 March 2011.

[167] Jones, Carl (2002). "Reptiles and Amphibians". *Handbook of ecological restoration: Principles of Restoration*. **2**. Cambridge University Press. p. 362. ISBN 0521791286.

[168] Towns, D.; Ballantine, W. (1993). "Conservation and restoration of New Zealand Island ecosystems". *Trends in Ecology & Evolution* **8**: 452. doi:10.1016/0169-5347(93)90009-E.

[169] Rauzon, Mark (2008). "Island restoration: Exploring the past, anticipating the future" (http://marineornithology.org/PDF/35_2/35_2_97-107.pdf). *Marine Ornithology* **35**: 97−107. .

[170] Diamond, Jared (1990). Towns, D; Daugherty, C; Atkinson, I. eds. *New Zealand as an archipelago: An international perspective* (http://192.206.154.93/upload/documents/science-and-technical/EcologicalRestorationNZIslands.pdf#page=8). Wellington: Conservation Sciences Publication No. 2. Department of Conservation. pp. 3−8. .

[171] PPP GDP estimates from different organisations vary. The International Monetary Fund's estimate is US$27,420. "Report for Selected Countries and Subjects" (http://www.imf.org/external/pubs/ft/weo/2010/02/weodata/weorept.aspx?sy=2008&ey=2015&scsm=1&ssd=1&sort=country&ds=.&br=1&pr1.x=30&pr1.y=9&c=196&s=PPPPC&grp=0&a=). International Monetary Fund. October 2010. . Retrieved 30 January 2011. The CIA World Factbook estimate is $28,000. "GDP − per capita (PPP)" (https://www.cia.gov/library/publications/the-world-factbook/rankorder/2004rank.html). The World Factbook, Central Intelligence Agency. . Retrieved 22 January 2011. The World Bank's estimate is US$29,352. "GDP per capita (current US$)" (http://web.archive.org/web/20110511123254/http://data.worldbank.org/indicator/NY.GDP.PCAP.CD). World Bank. . Retrieved 22 January 2011.

[172] "Currencies of the territories listed in the BS exchange rate lists" (http://www.bsi.si/en/financial-data.asp?MapaId=1239). Bank of Slovenia. . Retrieved 22 January 2011.

[173] "Human Development Index and components" (http://hdr.undp.org/en/media/HDR_2011_EN_Table1.pdf) (PDF). United Nations Development Programme. . Retrieved 3 January 2012.

[174] "2011 Index of Economic Freedom" (http://www.heritage.org/index/). The Heritage Foundation and Wall Street Journal. . Retrieved 15 January 2011.

[175] "NZ tops Travellers' Choice Awards" (http://www.stuff.co.nz/travel/396410). Stuff Travel. May 2008. . Retrieved 30 April 2010.

[176] "Historical evolution and trade patterns" (http://www.TeAra.govt.nz/en/1966/trade-external/1). *An Encyclopaedia of New Zealand*. 1966. . Retrieved 10 February 2011.

[177] Stringleman, Hugh; Peden, Robert (October 2009). "Sheep farming − Growth of the frozen meat trade, 1882−2001" (http://www.teara.govt.nz/en/sheep-farming/5/2). Te Ara − the Encyclopedia of New Zealand. . Retrieved 6 May 2010.

[178] Baker, John (February 2010) [originally published in 1966]. "Some Indicators of Comparative Living Standards" (http://www.TeAra. govt.nz/en/1966/standard-of-living/1/1). In McLintock, Alexander. *from An Encyclopaedia of New Zealand*. Te Ara – the Encyclopedia of New Zealand. . Retrieved 330 April 2010. Table pdf downloadable from (http://www.teara.govt.nz/files/ 3_308_StandardOfLiving_Comparison_0.pdf)

[179] Wilson, John (March 2009). "History – The later 20th century" (http://www.TeAra.govt.nz/en/history/6). Te Ara – the Encyclopedia of New Zealand. . Retrieved 2 February 2011.

[180] Nixon, Chris; Yeabsley, John (April 2010). "Overseas trade policy – Difficult times – the 1970s and early 1980s" (http://www.TeAra. govt.nz/en/overseas-trade-policy/5). Te Ara – the Encyclopedia of New Zealand. . Retrieved 22 January 2011.

[181] Evans, N.. "Up From Down Under: After a Century of Socialism, Australia and New Zealand are Cutting Back Government and Freeing Their Economies". *National Review* **46** (16): 47–51.

[182] Easton, Brian (November 2010). "Economic history – Government and market liberalisation" (http://www.TeAra.govt.nz/en/ economic-history/11). Te Ara – the Encyclopedia of New Zealand. . Retrieved 1 February 2011.

[183] Hazledine, Tim (1998) (PDF). *Taking New Zealand Seriously: The Economics of Decency* (http://www.ariplex.com/ ~economic-myth-busters/hazledine-taking nz seriously.pdf). HarperCollins Publishers. ISBN 1869502833. .

[184] "Unemployment" (http://www.socialreport.msd.govt.nz/paid-work/unemployment.html). 2010 Social report. . Retrieved 4 February 2011.

[185] Bingham, Eugene (7 April 2008). "The miracle of full employment" (http://www.nzherald.co.nz/business/news/article.cfm?c_id=3& objectid=10502512). *The New Zealand Herald*. . Retrieved 17 September 2008.

[186] "New Zealand Takes a Pause in Cutting Rates" (http://www.nytimes.com/2009/06/11/business/global/11nzrate.html). *The New York Times*. 10 June 2009. . Retrieved 30 April 2010.

[187] "New Zealand's slump longest ever" (http://news.bbc.co.uk/2/hi/business/8120196.stm). *BBC News*. 26 June 2009. . Retrieved 30 April 2010.

[188] Bascand, Geoff (February 2011). "Household Labour Force Survey: December 2010 quarter – Media Release" (http://web.archive.org/ web/20110429174323/http://www.stats.govt.nz/browse_for_stats/work_income_and_spending/employment_and_unemployment/ HouseholdLabourForceSurvey_MRDec10qtr.aspx). Statistics New Zealand. . Retrieved 4 February 2011.

[189] " Employment and Unemployment – June 2011 Quarter (http://www.dol.govt.nz/lmr/lmr-hlfs.asp)". NZ Department of Labour.

[190] Davenport, Sally (2004). "Panic and panacea: brain drain and science and technology human capital policy" (http://www.sciencedirect. com/science?_ob=ArticleURL&_udi=B6V77-4C007RT-1&_user=10&_coverDate=05/31/2004&_rdoc=6&_fmt=high&_orig=browse& _srch=doc-info(#toc#5835#2004#999669995#502989#FLA#display#Volume)&_cdi=5835&_sort=d&_docanchor=&_ct=9& _acct=C000050221&_version=1&_urlVersion=0&_userid=10&md5=37b8a5c81622b1687fbea5dfb51a0f38). *Research Policy* **33** (4): 617–630. doi:10.1016/j.respol.2004.01.006. .

[191] O'Hare, Sean (September 2010). "New Zealand brain-drain worst in world" (http://www.telegraph.co.uk/expat/expatnews/7973220/ New-Zealand-brain-drain-worst-in-world.html). *The Telegraph* (United Kingdom). .

[192] Collins, Simon (March 2005). "Quarter of NZ's brightest are gone" (http://www.nzherald.co.nz/nz/news/article.cfm?c_id=1& objectid=10114923). *New Zealand Herald*. .

[193] Winkelmann, Rainer (2000). "The labour market performance of European immigrants in New Zealand in the 1980s and 1990s". *The International Migration Review* (The Center for Migration Studies of New York) **33** (1): 33–58. doi:10.2307/2676011. JSTOR 2676011. Journal subscription required

[194] Bain 2006, p. 44.

[195] Groser, Tim (March, 2009). "Speech to ASEAN-Australia-New Zealand Free Trade Agreement Seminars" (http://www.beehive.govt.nz/ speech/speech-asean-australia-new-zealand-free-trade-agreement-seminars). New Zealand Government. . Retrieved 30 January 2011.

[196] "Improving Access to Markets:Agriculture" (http://www.mfat.govt.nz/Trade-and-Economic-Relations/NZ-and-the-WTO/ Improving-access-to-markets/0-agriculturenegs.php). New Zealand Ministry of Foreign Affairs and Trade. . Retrieved 22 January 2011.

[197] "New Zealand Economic and Financial Overview 2010: Industrial Structure and Principal Economic Sectors" (http://www.treasury.govt. nz/economy/overview/2010/09.htm). New Zealand Treasury. April 2010. . Retrieved 30 January 2011.

[198] O'Sullivan, Fran (April 2008). "Trade agreement just the start – Clark" (http://www.nzherald.co.nz/trade-deal-with-china/news/article. cfm?c_id=1501819&objectid=10502506&pnum=0). The New Zealand Herald. . Retrieved 30 April 2010.

[199] "China and New Zealand sign free trade deal" (http://www.nytimes.com/2008/04/07/business/worldbusiness/07iht-7tradefw. 11718461.html). *The New York Times*. April 2008. .

[200] "Key Tourism Statistics" (http://www.tourismresearch.govt.nz/Documents/Key Statistics/KeyTourismStatisticsApr2010.pdf) (PDF). Ministry of Tourism. April 2010. . Retrieved 30 April 2010.

[201] "International-Visitor-Arrivals Commentary" (http://www.tourismresearch.govt.nz/Data--Analysis/International-tourism/ International-Visitor-Arrivals/IVA-Commentary/). Tourismresearch. . Retrieved 20 January 2011.

[202] Easton, Brian (March 2009). "Economy – Agricultural production" (http://www.TeAra.govt.nz/en/economy/2). . Retrieved 22 January 2011.

[203] Stringleman, Hugh; Peden, Robert (March 2009). "Sheep farming – Changes from the 20th century" (http://www.TeAra.govt.nz/en/ sheep-farming/7). Te Ara – the Encyclopedia of New Zealand. . Retrieved 22 January 2011.

[204] Stringleman, Hugh; Scrimgeour, Frank (November 2009). "Dairying and dairy products – Dairying in the 2000s" (http://www.TeAra. govt.nz/en/dairying-and-dairy-products/10). Te Ara – the Encyclopedia of New Zealand. . Retrieved 22 January 2011.

[205] Stringleman, Hugh; Scrimgeour, Frank (March 2009). "Dairying and dairy products – Dairy exports" (http://www.TeAra.govt.nz/en/dairying-and-dairy-products/11). Te Ara – the Encyclopedia of New Zealand. . Retrieved 4 February 2011.

[206] "Global New Zealand – International Trade, Investment, and Travel Profile: Year ended June 2009 – Key Points" (http://www.stats.govt.nz/browse_for_stats/industry_sectors/imports_and_exports/global-nz-jun-09/key-points.aspx). Statistics New Zealand. June 2009. . Retrieved 4 February 2011.

[207] Stringleman, Hugh; Scrimgeour, Frank (March 2009). "Dairying and dairy products – Manufacturing and marketing in the 2000s" (http://www.TeAra.govt.nz/en/dairying-and-dairy-products/12). Te Ara – the Encyclopedia of New Zealand. . Retrieved 22 January 2011.

[208] Dalley, Bronwyn (March 2009). "Wine – The wine boom, 1980s and beyond" (http://www.TeAra.govt.nz/en/wine/6). Te Ara – the Encyclopedia of New Zealand. . Retrieved 22 January 2011.

[209] "Wine in New Zealand" (http://www.economist.com/node/10926423). *The Economist*. March 2008. .

[210] "Agricultural and forestry exports from New Zealand: Primary sector export values for the year ending June 2010" (http://www.maf.govt.nz/news-resources/statistics-forecasting/international-trade.aspx). New Zealand Ministry of Agriculture and Forestry. 14 January 2011. . Retrieved 8 April 2011.

[211] "Energy Data File 2009" (http://www.med.govt.nz/templates/MultipageDocumentTOC____41143.aspx). Ministry for Economic Development. July 2009. .

[212] "Frequently Asked Questions" (http://www.nzta.govt.nz/network/operating/faqs.html). New Zealand Transport Agency. . Retrieved 22 January 2011.

[213] "CIA – The World Factbook – New Zealand" (https://www.cia.gov/library/publications/the-world-factbook/geos/nz.html). . Retrieved 18 September 2009.

[214] Humphris, Adrian (April 2010). "Public transport – Passenger trends" (http://www.TeAra.govt.nz/en/public-transport/8). Te Ara – the Encyclopedia of New Zealand. . Retrieved 22 January 2011.

[215] Atkinson, Neill (November 2010). "Railways – Rail transformed" (http://www.TeAra.govt.nz/en/railways/11). Te Ara – the Encyclopedia of New Zealand. . Retrieved 22 January 2011.

[216] Atkinson, Neill (April 2010). "Railways – Freight transport" (http://www.TeAra.govt.nz/en/railways/6). Te Ara – the Encyclopedia of New Zealand. . Retrieved 22 January 2011.

[217] "International Visitors" (http://www.tourismresearch.govt.nz/Documents/International Market Profiles/Total Profile.pdf) (PDF). Ministry of Economic Development. June 2009. . Retrieved 30 January 2011.

[218] "10. Airports" (http://www.med.govt.nz/templates/MultipageDocumentPage____9038.aspx#P5641_412038). *Infrastructure Stocktake: Infrastructure Audit*. Ministry of Economic Development. December 2005. . Retrieved 30 January 2011.

[219] "Overview of the New Zealand Telecommunications Market 1987–1997" (http://www.med.govt.nz/templates/MultipageDocumentPage____4847.aspx). Ministry of Economic Development. November 2005. . Retrieved 30 January 2011.

[220] Budde, Paul. "New Zealand – Telecommunications – Major Players" (http://www.budde.com.au/Research/New-Zealand-Telecommunications-Major-Players.html). Budde Comm. . Retrieved 30 January 2011.

[221] "Estimated resident population of New Zealand" (http://www.stats.govt.nz/tools_and_services/tools/population_clock.aspx). Statistics New Zealand. . Retrieved 30 January 2011. The population clock updates every 10 minutes.

[222] "Subnational population estimates at 30 June 2009" (http://www.stats.govt.nz/methods_and_services/access-data/tables/subnational-pop-estimates.aspx). Statistics New Zealand. 30 June 2007. . Retrieved 30 April 2010.

[223] "Mercer 2010 Quality of Living survey highlights – Global" (http://www.mercer.com/referencecontent.htm?idContent=1128060). Mercer. May 2010. . Retrieved 30 April 2010.

[224] "Commentary" (http://www.stats.govt.nz/browse_for_stats/population/births/BirthsAndDeaths_HOTPDec09qtr/Commentary.aspx). *Births and Deaths: December 2009 quarter*. Statistics New Zealand. . Retrieved 27 April 2010.

[225] Department of Economic and Social Affairs Population Division (2009) (PDF). *World Population Prospects* (http://www.un.org/esa/population/publications/wpp2008/wpp2008_text_tables.pdf). 2008 revision. United Nations. . Retrieved 29 August 2009.

[226] "Ethnic groups in New Zealand" (http://www.stats.govt.nz/Census/2006CensusHomePage/QuickStats/quickstats-about-a-subject/culture-and-identity/ethnic-groups-in-new-zealand.aspx). *2006 Census QuickStats National highlights*. Statistics New Zealand. . Retrieved 18 January 2011.

[227] "Cultural diversity" (http://www.stats.govt.nz/Census/2006CensusHomePage/QuickStats/quickstats-about-a-subject/national-highlights/cultural-diversity.aspx). *2006 Census QuickStats National highlights*. Statistics New Zealand. . Retrieved 30 April 2010.

[228] When completing the census people could select more than one ethnic group (for instance, 53 percent of Māori identified solely as Māori, while the remainder also identified with one or more other ethnicities). "Māori Ethnic Population / Te Momo Iwi Māori" (http://www.stats.govt.nz/Census/2006CensusHomePage/QuickStats/quickstats-about-a-subject/maori/maori-ethnic-population-te-momo-iwi-maori.aspx). *QuickStats About Māori, Census 2006*. Statistics New Zealand. . Retrieved 30 April 2010.

[229] Collins, Simon (October 2010). "Ethnic mix changing rapidly" (http://www.nzherald.co.nz/nz/news/article.cfm?c_id=1&objectid=10678220). *New Zealand Herald*. .

[230] Dalby, Simon (1993). "The 'Kiwi disease': geopolitical discourse in Aotearoa/New Zealand and the South Pacific". *Political Geography* **12**: 437–988. doi:10.1016/0962-6298(93)90012-V.

[231] Callister, Paul (2004). "Seeking an Ethnic Identity: Is "New Zealander" a Valid Ethnic Category?" (http://panz.rsnz.org/wp-content/uploads/2010/01/nzpr-vol-30-1and-2_callister.pdf). *New Zealand Population Review* **30** (1&2): 5–22. .

[232] Misa, Tapu (8 March 2006). "Ethnic Census status tells the whole truth" (http://www.nzherald.co.nz/nz/news/article.cfm?c_id=1& objectid=10371473). *New Zealand Herald.* .

[233] "Draft Report of a Review of the Official Ethnicity Statistical Standard: Proposals to Address the 'New Zealander' Response Issue" (http:// web.archive.org/web/20101113182918/http://www.stats.govt.nz/~/media/Statistics/Methods and Services/Review ethnicity/ draft-reportof-reviewo-official-ethnicity-standard.ashx) (PDF). Statistics New Zealand. April 2009. . Retrieved 18 January 2011.

[234] Ranford, Jodie. "'Pakeha', Its Origin and Meaning" (http://maorinews.com/writings/papers/other/pakeha.htm). Māori News. . Retrieved 20 February 2008.

[235] Socidad Peruana de Medicina Intensiva (SOPEMI) (2000). *Trends in international migration: continuous reporting system on migration.* Organisation for Economic Co-operation and Development. pp. 276–278.

[236] Walrond, Carl (21 September 2007). "Dalmatians" (http://www.teara.govt.nz/en/dalmatians). Te Ara - the Encyclopedia of New Zealand. . Retrieved 30 April 2010.

[237] "New Zealand Peoples" (http://www.teara.govt.nz/en/new-zealand-peoples). Te Ara - the Encyclopedia of New Zealand. . Retrieved 30 April 2010.

[238] "International Migration Outlook – New Zealand 2009/10" (http://www.dol.govt.nz/publications/research/sopemi/2009-2010/ imo-2009-2010.pdf). New Zealand Department of Labour. 2010. p. 2. ISSN 1179-5085. . Retrieved 16 April 2011.

[239] "QuickStats About Culture and Identity: Birthplace and people born overseas" (http://www.stats.govt.nz/Census/ 2006CensusHomePage/QuickStats/quickstats-about-a-subject/culture-and-identity/birthplace-and-people-born-overseas.aspx). Statistics New Zealand. March 2006. . Retrieved 19 January 2011.

[240] For the percentages: "QuickStats About Culture and Identity – Birthplace and people born overseas" (http://www.stats.govt.nz/Census/ 2006CensusHomePage/QuickStats/quickstats-about-a-subject/culture-and-identity/birthplace-and-people-born-overseas.aspx). *2006 Census.* Statistics New Zealand. . Retrieved 30 April 2010.
For further detail within East Asia: "Culture and identity – Birthplace" (http://www.stats.govt.nz/methods_and_services/access-data/ TableBuilder/2006-census-pop-dwellings-tables/culture-and-identity/birthplace.aspx). *2006 Census Population and dwellings tables.* Statistics New Zealand. . Retrieved 30 April 2010.

[241] Butcher, Andrew; McGrath, Terry (2004). "International Students in New Zealand: Needs and Responses" (http://ehlt.flinders.edu.au/ education/iej/articles/v5n4/butcher/paper.pdf). *International Education Journal* **5** (4). .

[242] "QuickStats About Culture and Identity: Languages spoken" (http://www.stats.govt.nz/Census/2006CensusHomePage/QuickStats/ quickstats-about-a-subject/culture-and-identity/languages-spoken.aspx). Statistics New Zealand. March 2006. . Retrieved 20 February 2008.

[243] Hay, Maclagan & Gordon 2008, p. 14.

[244] Phillips, Jock (March 2009). "The New Zealanders – Bicultural New Zealand" (http://www.TeAra.govt.nz/en/the-new-zealanders/12). Te Ara – the Encyclopedia of New Zealand. . Retrieved 21 January 2011.

[245] "Māori Language Week – Te Wiki o Te Reo Maori" (http://www.nzhistory.net.nz/culture/maori-language-week). Ministry for Culture and Heritage. . Retrieved February 2008.

[246] Squires, Nick (May 2005). "British influence ebbs as New Zealand takes to talking Māori" (http://www.telegraph.co.uk/news/ worldnews/australiaandthepacific/newzealand/1490814/British-influence-ebbs-as-New-Zealand-takes-to-talking-Maori.html). *The Telegraph* (Great Britain). .

[247] "Waitangi Tribunal claim – Māori Language Week" (http://www.nzhistory.net.nz/culture/maori-language-week/ waitangi-tribunal-claim). Ministry for Culture and Heritage. July 2010. . Retrieved 19 January 2011.

[248] "Māori Television Launches 100 percent Māori Language Channel" (http://media.maoritelevision.com/default.aspx?tabid=211& pid=367). Māori Television. . Retrieved 30 April 2010.

[249] Of the 85,428 people that replied they spoke Samoan in the 2006 Census, 57,828 lived in the Auckland region.

[250] "2006 Census Data – QuickStats About Culture and Identity – Tables" (http://web.archive.org/web/20110610100515/http://www. stats.govt.nz/~/media/statistics/publications/census/2006-reports/quickstats-subject/culture-identity/ quickstats-about-culture-and-identity-tables.aspx) (XLS). *2006 Census.* Statistics New Zealand. . Retrieved 30 April 2010. In tables 28 (Religious Affiliation) and 19 (Languages Spoken by Ethnic Group)

[251] Languages listed here are those spoken by over 40,000 New Zealanders.

[252] New Zealand Sign Language Act 2006 No 18 (as at 30 June 2008), Public Act – New Zealand Legislation (http://www.legislation.govt. nz/act/public/2006/0018/latest/DLM372754.html). Legislation.govt.nz (2008-06-30). Retrieved on 2011-11-29.

[253] Dench, Olivia (July 2010). "Education Statistics of New Zealand: 2009" (http://www.educationcounts.govt.nz/publications/ece/2507/ 80221). Education Counts. . Retrieved 19 January 2011.

[254] Tertiary education in New Zealand is used to describe all aspects of post-school education and training. Its ranges from informal non-assessed community courses in schools through to undergraduate degrees and advanced, research-based postgraduate degrees.

[255] "Education Act 1989 No 80 (as at 01 February 2011), Public Act. Part 14: Establishment and disestablishment of tertiary institutions, Section 62: Establishment of institutions" (http://www.legislation.govt.nz/act/public/1989/0080/latest/DLM183668.html). *Education Act 1989 No 80.* New Zealand Parliamentary Counsel Office/Te Tari Tohutohu Pāremata. 1 February 2011. . Retrieved 15 August 2011.

[256] "Studying in New Zealand: Tertiary education" (http://www.nzqa.govt.nz/studying-in-new-zealand/tertiary-education). New Zealand Qualifications Authority. . Retrieved 15 August 2011.

[257] "Educational attainment of the population" (http://www.educationcounts.govt.nz/__data/assets/excel_doc/0007/17836/ Education_attainment_of_the_population.xls) (xls). Education Counts. 2006. . Retrieved 21 February 2008.

[258] "QuickStats About Culture and Identity: Religious affiliation" (http://www.stats.govt.nz/Census/2006CensusHomePage/QuickStats/ quickstats-about-a-subject/culture-and-identity/religious-affiliation.aspx). Statistics New Zealand. . Retrieved 20 January 2011.

[259] Another 6 percent objected to stating their religion. Statistics NZ do not report a total percentage for "Other" religions. Depending on how many people claimed both Christian and other religions, this could range from 3 to 5 percent. These percentages are based on the usually resident population, excluding another 7 percent of people who did not provide usable information.

[260] "Quick Stats About culture and Identity— 2006 Census" (http://web.archive.org/web/20110610100533/http://www.stats.govt.nz/~/ media/Statistics/Publications/Census/2006-reports/quickstats-subject/Culture-Identity/qstats-about-culture-and-identity-2006-census.pdf) (PDF). Statistics New Zealand. . Retrieved 28 September 2007.

[261] "Subnational population estimates" (http://www.stats.govt.nz/browse_for_stats/population/estimates_and_projections/ subnational-pop-estimates-tables.aspx). Statistics New Zealand. . Retrieved 2 November 2010.

[262] Kennedy 2007, p. 398.

[263] Hearn, Terry (March 2009). "English – Importance and influence" (http://www.TeAra.govt.nz/en/english/). Te Ara – the Encyclopedia of New Zealand. . Retrieved 21 January 2011.

[264] "Conclusions – British and Irish immigration" (http://www.nzhistory.net.nz/culture/home-away-from-home/conclusions). Ministry for Culture and Heritage. March 2007. . Retrieved 21 January 2011.

[265] Stenhouse, John (November 2010). "Religion and society – Māori religion" (http://www.TeAra.govt.nz/en/religion-and-society/4). Te Ara – the Encyclopedia of New Zealand. . Retrieved 21 January 2011.

[266] "Māori Social Structures" (http://www.justice.govt.nz/publications/publications-archived/2001/ he-hinatore-ki-te-ao-maori-a-glimpse-into-the-maori-world/part-1-traditional-maori-concepts/maori-social-structures). Ministry of Justice. March 2001. . Retrieved 21 January 2011.

[267] Kennedy 2007, p. 400.

[268] Kennedy 2007, p. 399.

[269] Phillips, Jock (March 2009). "The New Zealanders – Post-war New Zealanders" (http://www.TeAra.govt.nz/en/the-new-zealanders/ 10). Te Ara – the Encyclopedia of New Zealand. . Retrieved 21 January 2011.

[270] Phillips, Jock (March 2009). "The New Zealanders – Ordinary blokes and extraordinary sheilas" (http://www.TeAra.govt.nz/en/ the-new-zealanders/11). Te Ara – the Encyclopedia of New Zealand. . Retrieved 21 January 2011.

[271] Phillips, Jock (March 2009). "Rural mythologies – The cult of the pioneer" (http://www.TeAra.govt.nz/en/rural-mythologies/5). Te Ara – the Encyclopedia of New Zealand. . Retrieved 21 January 2011.

[272] Swarbrick, Nancy (June 2010). "Creative life – Visual arts and crafts" (http://www.TeAra.govt.nz/en/creative-life/2). Te Ara – the Encyclopedia of New Zealand. . Retrieved 4 February 2011.

[273] McLintock, Alexander, ed (April 2009) [originally published in 1966]. "Elements of Carving" (http://www.TeAra.govt.nz/en/1966/ maori-art/4). *from An Encyclopaedia of New Zealand*. Te Ara – the Encyclopedia of New Zealand. . Retrieved 15 February 2011.

[274] McLintock, Alexander, ed (April 2009) [originally published in 1966]. "Surface Patterns" (http://www.TeAra.govt.nz/en/1966/ maori-art/5). *from An Encyclopaedia of New Zealand*. Te Ara – the Encyclopedia of New Zealand. . Retrieved 15 February 2011.

[275] McKay, Bill (2004). "Māori architecture: transforming western notions of architecture" (http://www.library.uq.edu.au/ojs/index.php/ fab/article/viewFile/108/126). *Fabrications: the Journal of the Society of Architectural Historians, Australia and New Zealand* **14** (1&2): 1–12. .

[276] McLintock, Alexander, ed (April 2009) [originally published in 1966]. "Painted Designs" (http://www.TeAra.govt.nz/en/1966/ maori-art/8). *from An Encyclopaedia of New Zealand*. Te Ara – the Encyclopedia of New Zealand. . Retrieved 15 February 2011.

[277] McLintock, Alexander, ed (April 2009) [originally published in 1966]. "Tattooing" (http://www.TeAra.govt.nz/en/1966/maori-art/9). *from An Encyclopaedia of New Zealand*. Te Ara – the Encyclopedia of New Zealand. . Retrieved 15 February 2011.

[278] "Beginnings – history of NZ painting" (http://www.nzhistory.net.nz/culture/nz-painting-history/beginnings). Ministry for Culture and Heritage. December 2010. . Retrieved 17 February 2011.

[279] "A new New Zealand art – history of NZ painting" (http://www.nzhistory.net.nz/culture/nz-painting-history/a-new-new-zealand-art). Ministry for Culture and Heritage. November 2010. . Retrieved 16 February 2011.

[280] "Contemporary Maori art" (http://www.nzhistory.net.nz/culture/nz-painting-history/contemporary-maori-art). Ministry for Culture and Heritage. November 2010. . Retrieved 16 February 2011.

[281] Rauer, Julie. "Paradise Lost: Contemporary Pacific Art At The Asia Society" (http://www.asianart.com/exhibitions/paradise/article. html). Asia Society and Museum. . Retrieved 17 February 2011.

[282] McLintock, Alexander, ed (April 2009) [originally published in 1966]. "Textile Designs" (http://www.TeAra.govt.nz/en/1966/ maori-art/10). *from An Encyclopaedia of New Zealand*. Te Ara – the Encyclopedia of New Zealand. . Retrieved 15 February 2011.

[283] Keane, Basil (March 2009). "Pounamu – jade or greenstone – Implements and adornment" (http://www.TeAra.govt.nz/en/ pounamu-jade-or-greenstone/4). Te Ara – the Encyclopedia of New Zealand. . Retrieved 17 February 2011.

[284] Wilson, John (March 2009). "Society – Food, drink and dress" (http://www.TeAra.govt.nz/en/society/9). Te Ara – the Encyclopedia of New Zealand. . Retrieved 17 February 2011.

[285] Swarbrick, Nancy (June 2010). "Creative life – Design and fashion" (http://www.TeAra.govt.nz/en/creative-life/3). Te Ara – the Encyclopedia of New Zealand. . Retrieved 22 January 2011.

[286] "Fashion in New Zealand – New Zealand's fashion industry" (http://www.economist.com/displayStory. cfm?Story_ID=E1_TDSGGNTD). *The Economist*. 28 February 2008. . Retrieved 6 August 2009.

[287] Swarbrick, Nancy (June 2010). "Creative life – Writing and publishing" (http://www.TeAra.govt.nz/en/creative-life/6). Te Ara – the Encyclopedia of New Zealand. . Retrieved 22 January 2011.

[288] "The making of New Zealand literature" (http://www.nzhistory.net.nz/culture/literature-in-new-zealand-1930-1960). Ministry for Culture and Heritage. November 2010. . Retrieved 22 January 2011.

[289] "New directions in the 1930s – New Zealand literature" (http://www.nzhistory.net.nz/culture/literature-1940-60/1930s). Ministry for Culture and Heritage. August 2008. . Retrieved 12 February 2011.

[290] "The war and beyond – New Zealand literature" (http://www.nzhistory.net.nz/culture/nz-literature/the-growth-of-publishing). Ministry for Culture and Heritage. November 2007. . Retrieved 12 February 2011.

[291] Swarbrick, Nancy (June 2010). "Creative life – Music" (http://www.TeAra.govt.nz/en/creative-life/7). Te Ara – the Encyclopedia of New Zealand. . Retrieved 21 January 2011.

[292] McLintock, Alexander, ed (April 2009) [originally published in 1966]. "Maori Music" (http://www.TeAra.govt.nz/en/1966/maori-music/1). *from An Encyclopaedia of New Zealand*. Te Ara – the Encyclopedia of New Zealand. . Retrieved 15 February 2011.

[293] McLintock, Alexander, ed (April 2009) [originally published in 1966]. "Musical Instruments" (http://www.TeAra.govt.nz/en/1966/maori-music/6). *from An Encyclopaedia of New Zealand*. Te Ara – the Encyclopedia of New Zealand. . Retrieved 16 February 2011.

[294] McLintock, Alexander, ed (April 2009) [originally published in 1966]. "Instruments Used for Non-musical Purposes" (http://www.TeAra.govt.nz/en/1966/maori-music/7). *from An Encyclopaedia of New Zealand*. Te Ara – the Encyclopedia of New Zealand. . Retrieved 16 February 2011.

[295] McLintock, Alexander, ed (April 2009) [originally published in 1966]. "Music: General History" (http://www.TeAra.govt.nz/en/1966/music/1). *from An Encyclopaedia of New Zealand*. Te Ara – the Encyclopedia of New Zealand. . Retrieved 15 February 2011.

[296] McLintock, Alexander, ed (April 2009) [originally published in 1966]. "Music: Brass Bands" (http://www.TeAra.govt.nz/en/1966/music/3). *from An Encyclopaedia of New Zealand*. Te Ara – the Encyclopedia of New Zealand. . Retrieved 14 April 2011.

[297] McLintock, Alexander, ed (April 2009) [originally published in 1966]. "Music: Pipe Bands" (http://www.TeAra.govt.nz/en/1966/music/7). *from An Encyclopaedia of New Zealand*. Te Ara – the Encyclopedia of New Zealand. . Retrieved 14 April 2011.

[298] Swarbrick, Nancy (June 2010). "Creative life – Performing arts" (http://www.TeAra.govt.nz/en/creative-life/8). Te Ara – the Encyclopedia of New Zealand. . Retrieved 21 January 2011.

[299] Swarbrick, Nancy (June 2010). "Creative life – Film and broadcasting" (http://www.TeAra.govt.nz/en/creative-life/5). Te Ara – the Encyclopedia of New Zealand. . Retrieved 21 January 2011.

[300] Cieply, Michael; Rose, Jeremy (October 2010). "New Zealand Bends and 'Hobbit' Stays" (http://www.nytimes.com/2010/10/28/business/media/28hobbit.html). *New York Times*. .

[301] "Production Guide: Locations" (http://www.filmnz.com/production-guide/locations.html). Film New Zealand. . Retrieved 21 January 2011.

[302] "Only peace protects freedoms in post-9/11 world" (http://en.rsf.org/only-peace-protects-freedoms-in-22-10-2008,29031). Reporters Without Borders. 22 October 2008. . Retrieved 30 April 2010.

[303] Hearn, Terry (March 2009). "English – Popular culture" (http://www.TeAra.govt.nz/en/english/12). Te Ara – the Encyclopedia of New Zealand. . Retrieved 22 January 2022.

[304] Phillips, Jock (February 2011). "Sports and leisure – Organised sports" (http://www.TeAra.govt.nz/en/sports-and-leisure/4). Te Ara – the Encyclopedia of New Zealand. . Retrieved 23 March 2011.

[305] Crawford, Scott (January 1999). "Rugby and the Forging of National Identity" (http://www.la84foundation.org/SportsLibrary/ASSHSSH/ASSHSSH11.pdf). In Nauright, John. *Sport, Power And Society In New Zealand: Historical And Contemporary Perspectives*. ASSH Studies In Sports History. .

[306] Fougere, Geoff (1989). "Sport, culture and identity: the case of rugby football" (http://www.cabdirect.org/abstracts/19901879245.html;jsessionid=9057684DAADD50F83FAE7254DC066545). In Novitz, David; Willmott, Bill. *Culture and identity in New Zealand*. pp. 110–122. ISBN 0-477-01422-4. .

[307] "Rugby, racing and beer" (http://www.nzhistory.net.nz/media/photo/rugby-racing-and-beer). Ministry for Culture and Heritage. August 2010. . Retrieved 22 January 2011.

[308] Derby, Mark (December 2010). "Māori–Pākehā relations – Sports and race" (http://www.TeAra.govt.nz/en/maori-pakeha-relations/4). Te Ara – the Encyclopedia of New Zealand. . Retrieved 4 February 2011.

[309] "ABS medal tally: Australia finishes third" (http://www.abs.gov.au/Ausstats/abs@.nsf/57a31759b55dc970ca2568a1002477b6/be9f47591541e29eca256ef40004f25a!OpenDocument). Australian Bureau of Statistics. 30 August 2004. . Retrieved 17 February 2008.

[310] Bain 2006, p. 69.

[311] "World mourns Sir Edmund Hillary" (http://news.theage.com.au/national/world-mourns-sir-edmund-hillary-20080111-1ldx.html). *The Age* (Australia). January 2008. .

[312] "Sport and Recreation Participation Levels" (http://www.activenzsurvey.org.nz/Documents/Participation-Levels.pdf). Sport and Recreation New Zealand. 2009. . Retrieved 30 April 2010.

[313] Yousef, Robyn (January 2011). "Waka ama: Keeping it in the family" (http://www.nzherald.co.nz/maori/news/article.cfm?c_id=252&objectid=10703178). *New Zealand Herald*. .

References

Bibliography

- Bain, Carolyn (2006). *New Zealand*. Lonely Planet. ISBN 1741045355.
- Garden, Donald (2005). Stoll, Mark. ed. *Australia, New Zealand, and the Pacific: an environmental history*. Nature and Human Societies. ABC-CLIO/Greenwood. ISBN 9781576078686.
- Kennedy, Jeffrey (2007). "Leadership and Culture in New Zealand". In Chhokar, Jagdeep; Brodbeck, Felix; House, Robert. *Culture and Leadership Across the World: The GLOBE Book of In-Depth Studies of 25 Societies*. US: Psychology Press. ISBN 978-0-8058-5997-3.
- Hay, Jennifer; Maclagan, Margaret; Gordon, Elizabeth (2008). *Dialects of English: New Zealand English*. Edinburgh University Press. ISBN 9780748625291.
- King, Michael (2003). *The Penguin History of New Zealand*. New Zealand: Penguin Books. ISBN 9780143018674.
- Mein Smith, Philippa (2005). *A Concise History of New Zealand*. Australia: Cambridge University Press. ISBN 0521542286.

Further reading

- Bateman, David, ed (2005). *Bateman New Zealand Encyclopedia* (6th ed.). ISBN 1869536010.
- Sinclair, Keith; revised by Dalziel, Raewyn (2000). *A History of New Zealand*. ISBN 9780140298758.
- Statistics New Zealand. *New Zealand Official Yearbook* (annual). ISBN 1869537769 (2010).

External links

Government

- New Zealand Government portal (http://newzealand.govt.nz/)
- Ministry for Culture and Heritage (http://www.mch.govt.nz/) – includes information on flag, anthems and coat of arms
- Statistics New Zealand (http://www.stats.govt.nz/)

Travel

- New Zealand travel guide from Wikitravel
- Tourism New Zealand (http://www.newzealand.com/)

Other

- Te Ara, The Encyclopedia of New Zealand (http://www.teara.govt.nz/)
- NZHistory.net.nz New Zealand history website (http://www.nzhistory.net.nz/)
- New Zealand (http://ucblibraries.colorado.edu/govpubs/for/newzealand.htm), directory from *UCB Libraries GovPubs*
- New Zealand weather (http://www.metservice.co.nz/)

kbd:Зелэнд ЩІэ gag:Eni Zelandiya mrj:У Зеланди rue:Новый Зеланд

Article Sources and Contributors

Sportavex *Source*: http://en.wikipedia.org/w/index.php?title=Sportavex *Contributors*: Ardfern, Djln, Edward the Confessor, Gimboid13, Kotukunui, Mikepaauwe

Homebuilt_aircraft *Source*: http://en.wikipedia.org/w/index.php?title=Homebuilt_aircraft *Contributors*: AMCKen, Ahunt, Alexmcfire, Apirkle, Appraiser, Are david, Arpingstone, Arrivisto, BACbKA, CBorges, CanisRufus, Casmeli, Chairboy, Collykreidler, D6, DOHC Holiday, DexDor, Djapa84, Eaapilot, Ericg, EthanL, Eurosong, Falconus, Fletcher, Frank Lofaro Jr., Frazzydee, Gbleem, Geoffrey Wickham, Gmcneel, Hmains, IanOsgood, Joel7687, JohnPiepers, Justsoaring, KVDP, Karl Dickman, Kirill Borisenko, Krellis, Leszek Jańczuk, Lightmouse, Male1979, MarsRover, Martylunsford, MilborneOne, Mjs072, Moggy58, Nimbus227, Optimist on the run, Peteotube, Pik-26, Psb777, RicReis, Rlandmann, Robert Merkel, Rushikeshjoshi003, Rv8, Rvator, Sardanaphalus, Shreditor, Siafu, Skipper2, SkipperPilot, SwiftAircraft, The PIPE, Tobias Bergemann, Trekphiler, Unixxx, Vegaswikian, WTucker, Wapati8, Wavelength, Wingman4l7, Woohookitty, Zenswashbuckler, 94 anonymous edits

Aviation *Source*: http://en.wikipedia.org/w/index.php?title=Aviation *Contributors*: 16@r, 999xyzma, A Macedonian, A little insignificant, AGToth, Acidplasticman, Aeronews, Aesopos, Agentblank, Ahmad Fauzi HMS, Ahunt, Airplaneman, Akradecki, Altenmann, Andy Dingley, Antandrus, AntonioMartin, Arpingstone, Ashton1983, Astonfw, Azul247, BarretB, BatteryIncluded, Beetstra, Bensin, Bethaso, Bidouilleur, Binksternet, Blake-, Blimpguy, Blueshade, BokicaK, Braisim, Broocrew, Bsadowski1, CJ, CLAES, Caltas, CambridgeBayWeather, Can't sleep, clown will eat me, Capricorn42, CaptainVindaloo, Carltzau, Carly88, Charlemain, CharlieRCD, Chase me ladies, I'm the Cavalry, Chasingsol, Cheeser1, Cherubino, Chrhardy, Chris the speller, Civil Engineer III, Cleared as filed, Coastwise, Courcelles, CultureDrone, Cyrusc, D6, Dancter, Danio, DavidLevinson, Dbchip, Deflectico, Delldot, Deltaker, DexDor, Docboat, Dolphin51, Dpm64, Durksteel, Durova, Dzenanz, EH101, EdJogg, Edivorce, Edward, EricSkinner, Evil Merlin, FMB, Fan-1967, Favonian, Felixboy, Fieldday-sunday, Flymeoutofhere, GRAHAMUK, Galoubet, Generaleskimo, Genius101, Gilliam, Glane23, Glenn, Gralo, Greyengine5, Grim23, Gun Powder Ma, Gundersen53, HJ Mitchell, Halo13245, Headbomb, Heltzen, Hmains, Holwil, Hsrc, Huji, Hulagutten, Ictglen, Idleguy, Iespindia, Imarquecm, ImperatorExercitus, Incidious, Inter16, Itzalwaysme, JColgan, JForget, JUtpadel, JackSparrow Ninja, Jagged 85, Jagoboy, Jaho, Jamescope, Jan Pospíšil, Jeffeh, Jefferry, Jerelabs, Jghiii, Jj137, Jj98, Jmcc150, Joe d'Eon, Joefaust, JoelKatz, John, John254, Joshp151, Jusdafax, KF, Kaisershatner, Kenny Moens, Kes321, Key45, Kingpin13, Kobek81, Koplimek, Kosebamse, Kovy999, Krellis, Kristian Finn, Ksyrie, Lahiru k, Leboman, Leopardjferry, Lightmouse, LogicUndefied, Look it hobo, Loren36, LovesMacs, Lupinelawyer, M-le-mot-dit, Macedonian, Mani1, MarSch, Marine 69-71, Marsian, Marskuzz, Martijn Hoekstra, Martpol, MarylandArtLover, MastrPlanner, Matt thomas, Matthewinventor, Maury Markowitz, Mav, McSly, Mcmxl, Mike Rosoft, Mike Stramba, MilborneOne, Moondoll, Mrclean9865, Mrg3105, Mroach, Ms2ger, Nascar1996, Nilmerg, Novum, NuclearWarfare, OMGsplosion, OrgasGirl, Ottre, Oxymoron83, Paj1113, Patoune08, Patrick, Persian Poet Gal, Peter Bankss, Philip Trueman, Phydeaux, PilotFrazee, Pingveno, Plane Person, Planedriver27, Plenumchamber, PseudoSudo, Pvtflyer, Pyrrhus16, Quintote, Qwitchibo, R'n'B, RJHall, RandomP, Rausch, Ray Trygstad, RemiOo, Rgvis, Rich Farmbrough, RichardF, Richardj311, RightSideNov, RivGuySC, Rjwilmsi, Rodrigo Cornejo, Rokus2000, Rror, Rurp, SD5, SDC, SQGibbon, Sardanaphalus, SaxTeacher, Seaphoto, Sgfaig, Sharkface217, Sherbrooke, SilkTork, Siroxo, Slakr, SmilesALot, Snowmanradio, Snowolf, Soladee, Spiress, Splette, SpuriousQ, Steelpillow, Stemonitis, Stwalkersock, Sue Rangell, Suffusion of Yellow, Supercowfan, Superdudeonfire, Syclop, Synchronism, Tassedethe, TastyPoutine, Tatrgel, TatyaM, Tennisuser123, TestPilot, Th1rt3en, The Rationalist, The Red, TheParanoidOne, Thedjatclubrock, Theguru320, Them308, Themfromspace, Tide rolls, Tkpowell, Tobby72, TopTopView, Toprohan, Trademark123, Treesmill, Trevor MacInnis, TruckCard, Unioneagle, Unused0029, Van der Hoorn, Vegaswikian, Vera.Dvoinos, Vgy7ujm, Vicki Rosenzweig, WJBscribe, WLU, Wangi, Windchaser, Wolfkeeper, Woohookitty, Xp54321, Yamla, Zocky, Zoke, Zonk43, Zzuuzz, Žiedas, 508 anonymous edits

Air_show *Source*: http://en.wikipedia.org/w/index.php?title=Air_show *Contributors*: 790, Acdx, Ageekgal, Aircraftmac, Airodyssey, Airshow4444, Akradecki, Alpertonian, Amberrock, AnnaFrance, Antheii, Anthony Appleyard, Apocno10, ArmanJan, Arpingstone, Atoc-tob, BabyNuke, Baldtaco, Balloonguy, Bankhallbretherton, Bbmav, Belovedfreak, Bennerg, BilCat, Bobblewik, Bobo192, Borgx, Brian Crawford, Bryan Derksen, Bzuk, CWRamey, Capt. Yaw, CarolGray, Catabv23, CharlesC, Chase me ladies, I'm the Cavalry, Chestergunn, ChrisW, Conversion script, Correctemundo, CountyCalendar, Crazycomputers, D2180s, Dannysoong, Dave-ros, David.Monniaux, DeAceShooter, Decimal10, Degen Earthfast, Dimitrii, Dolovis, Dralwik, Edward, ElvisFan1967, Eponymous one, FClef, FJM, Filtv, Flightlineuk, Florentino floro, Fyyer, Gaius Cornelius, Gangasudhan, Georgejdorner, Geotec22, Guliolopez, Hairuleff, Highflying, Hmains, Hu12, Huaiwei, Hydrargyrum, ILovePlankton, Implificator, J-Star, Jammers007, Jim G. Smtih, Jjron, JohnI, Jordiferrer, Jowensii, Joyradost, Julia W, Karlkoeppen, KevinCuddeback, Koavf, Kotukunui, Kross, Ksyrie, Kubanczyk, Kurtkratchman, Kznf, Lexusuns, Littleteddy, LuckyThracian, MHBowden, MVPZidane, Mac, Magnus Manske, Makemi, Malfita, Marianello, Mark83, Markdom, Mav, Maximilian Schönherr, Mbz1, McSly, Mdd4696, Michael Devore, Michael Zimmermann, MilborneOne, Militaryairshows, Mr.Z-man, Muad, Nagra11, Neet nl, Nigel Ish, NotACow, Nsaa, Nv8200p, PRRfan, Pax:Vobiscum, Paxsimius, Petersam, Planepull3, PlusMinus, Postdlf, Psb777, QueenCake, Raven4x4x, Retired username, ReyBrujo, Rich Farmbrough, Richard Arthur Norton (1958-), Rje, Rklawton, Rlandmann, Roke, Rsduhamel, Sam Blacketer, SchuminWeb, Severnapark, Shinerunner, SiobhanHansa, Skyflash, Sniperz11, Southendnick, Sp33dyphil, Spencer, Starionwolf, Steelerdon, SteinbDJ, Stephangerlach, Steven J. Anderson, Stocksfan, Taasss, Tannin, TestPilot, The Rambling Man, Thecreatorrulz, Themeweaver, Trevor MacInnis, Tylko, Ukar69, Vegaswikian, Veinor, Web kai2000, Wildcat dunny, Will Beback, Willster3, Woohookitty, Yogi de, Yuriybrisk, ZapThunderstrike, Zeimusu, Zscout370, Éire32, 256 anonymous edits

Paraparaumu *Source*: http://en.wikipedia.org/w/index.php?title=Paraparaumu *Contributors*: Alan Liefting, AmiDaniel, Andy king50, Annarchy, Axver, Ben Arnold, Birdhurst, Bobo192, Darius Dhlomo, Divinecirinde, Enzedbrit, GCFreak2, Gadfium, Galoubet, Gogo Dodo, Grutness, Howbizareay, Hugo999, Jasonlvc, Jeff79, Kuru, Kwamikagami, LJ Holden, LarryTheGreat, Mendors, Micknz, NevilleDNZ, Nzfauna, O, Padness, Paraparam Bigot, Poccil, Reedy, Steverwanda, Thesupernaut, Vardion, Will Beback, XLerate, 88 anonymous edits

Matamata *Source*: http://en.wikipedia.org/w/index.php?title=Matamata *Contributors*: ALargeElk, Ahkitj, Alan Liefting, Aliosos, Arria Belli, Beast from the Bush, Burek, Catchpolejames, Cgoodwin, Cmdrjameson, Crusadeonilliteracy, D3, Dar-Ape, Dramatic, Eranah, Eric119, Fuzheado, Gadfium, Grutness, Happyfeijoa, Helenalex, Ingolfson, Jj137, Jodiandalan, Kaimai, Kaiserm, Katetirau, Lcmortensen, Lemoninacup, Loganberry, Lt, Macronencer, Malathos, Michellecrisp, Mike subritzky, Missbambi, Noca2plus, Nurg, Nzseries1, Onco p53, Ot, Penfald, Pengo, Popsracer, Pscone, Pwntpwntpwnt, Quadell, Qwyrxian, Ryan4314, Schwede66, Sionus, Squids and Chips, Steinsky, Thebrunners, Toll25, Valley2city, Waikatopakeha, West Brom 4ever, Whkoh, Windofkeltia, XLerate, Yms, 126 anonymous edits

Tauranga *Source*: http://en.wikipedia.org/w/index.php?title=Tauranga *Contributors*: 1908ben, A. Carty, Ahkitj, Alansohn, Ale jrb, Allstarecho, Amdtelrunya, Amy231191, Andkore, Andrewpmk, Aranel, Bandhclassic, Beans42, Bearcat, Ben Arnold, Biross, Bmcdonaldnz, Bmnz, Bozzio, Brian, Burnie7, Buttsta, Buzz Nicholson, Can't sleep, clown will eat me, CheDP, Cmdrjameson, CommonsDelinker, Courcelles, David Kernow, Deflective, Denham.cook, DerbyCountyinNZ, DianeBlackmore, DI2000, Drakehellman, ESarge, Ecokiwi, Ed Kruger, Enzedbrit, EoGuy, FRED, FieldMarine, Frmatt, Gadfium, Gogo Dodo, Graham87, Gregfitzy, Grstain, Grutness, Gwarm, Haleereid, Hazyd, Hugo999, Ingolfson, J.delanoy, Jeromefarrell, Jhendin, Kahane6969, Kahnage, Kaiwhakahaere, Kappa, Kasatali, Kbh3rd, Kevymtnz, Khandallah21, Khoikhoi, Kotukunui, Kungming2, Kwamikagami, Lcmortensen, Leuko, LibStar, Lightmouse, Lilcamel, MER-C, MJCdetroit, Maias, Mamndassan, Mathmo, Matt von Furrie, Mattl2001, Mmgdertt, Monkey32, Moriori, Mouseluva, MozzazzoM, Nk, Noah Salzman, Nurg, Nz kiwi gal monkey, Oglerman, Onco p53, Orderinchaos 2, Papamoa8, Paulypaul78, Perfecto, Phil McKenzie, Plump thunder, Pootangalang, Popsracer, Pumpmeup, QFSE Media, Qupada, Random Passer-by, Reenem, Rich Farmbrough, Rocastelo, Rui Gabriel Correia, Runnermonkey, Sam Hocevar, Sam Tauranga, Scbarry, Schwede66, Seaphoto, Shedcat, ShelfSkewed, Shellzzy, SimonP, Skyring, SlackerMom, Spartan-James, Stacy.raymond, Surfnz, TJRB, TRBP, Tayste, Teggles, TerriersFan, Thebluelagoon, Tristanb, Useight, Videomaniac29, Viletraveller, Waimarino36, Wazza68, WikiDon, XLerate, Zigger, Æk, 310 anonymous edits

New_Zealand *Source*: http://en.wikipedia.org/w/index.php?title=New_Zealand *Contributors*: (, +++++CS+++++, -- April, -unicycle-pro-, 041744, 100110100, 10speed, 123username, 130.94.122.xxx, 1YUI0P, 1brettsnyder, 209.179.207.xxx, 23dazed, 23prootie, 24.189.53.xxx, 28421u2232nfenfcenc, 2sc945, 334a, 40010, 5 albert square, 565dj, 77hydro77, 9Nak, A. di M., A.J.Chesswas, A2raya07, ACSE, ADMBAIRD, AJD, AKMask, Abdullah@xtra.co.nz, Abovemost, Abrech, AbruCH, Acalamari, Accounting4Taste, AceNZ, Achangeisasgoodasa, Acntx, Acroterion, Adabow, Adamk, Adamross, Addshore, Aeon1006, Agentbikes, Ahkitj, Ahoerstemeier, Aircorn, Ajaxrools, Akanemoto, Akonga, Alan Liefting, AlanD, Alansohn, Alexandre.dussault, AlexiusHoratius, Alexjgunn, Alexkorn, Alexy527, Alfonc330, Algont, Alinor, Allstarecho, Alphaboi867, Alphachimp, Altenmann, Amakuru, Amaste, Amazong, Amerikiwi, Amgine, Amgine315, Amiyagenius, Amrush, Anand Karia, Andkore, Andre Engels, Andrew D White, Andrew Levine, Andrew Yong, AndrewHowse, Andrewlp1991, Andrewpmk, Andromeda321, Andrwsc, Andy Marchbanks, Andydrew1234, Andyjsmith, Anger22, Angmering, Angr, Animum, Ankur Chattopadhyay, Anna Lincoln, Annarchy, Annonymous, Anonymous Dissident, Anonymous editor, Anshuman.jrt, AntaineNZ, Antandrus, Antonio Lopez, Aotearoadub, ApprenticeFan, Arashi, Archer3, Archienz, ArchonMeld, Arctic Night, Arctic.gnome, ArdentV, Ardt, Arguss, Aridd, ArielGold, Arjun01, Armada579, Arnold08, Arpingstone, Arria Belli, Art LaPella, Arx Fortis, Arybella, Ascidian, Asdf333333, Ashwinosoft, AssegaiAli, Astral, AstroHurricane001, Astrotrain, Atarr, Atlantis Hawk, Atoric, Attilios, AuburnPilot, Audacitor, Aude, Augen Zu, Aum108, Auric, Ausierula33, AussieDownUnder, AussieLegend, Avala, Avenue, Avicennasis, Avraham, AxG, Axxgreazz, Az1568, Azazyel, AzureCitizen, B-stefpe, B. van der Wee, BD2412, Backspace, Badenoch, Baiji, BaileyDCampbell, Balastda, Balsa10, Balthazar, Barefootguru, Barek, Baronnet, Barryob, Bastin, Bazonka, Bbatsell, Bbert1994, Bbx, Bearcat, Bearly541, Beatles5445, Before My Ken, Belligero, Ben Arnold, Ben Ben, Ben whitehouse12, Ben-Zin, Benandorsqueaks, Bencannon, Bender235, Benky, Benzen, Beonid, Berlinerzeitung, Bfigura's puppy, Bgtd, Bhadani, Bheinola, Bidabadi, Big Adamsky, Big Brother 1984, BigBadaboom0, BilboBaggins182, Billthesinglingmonkey, Billymac00, Birdhurst, Birdman1, Biruitorul, Biss313, Bjdehut, Bjelleklang, Bkell, Bkonrad, Blander, Blarvink, BlazingSpeed, Bletch, BloodDoll, BlueLankan, Bluelion, BobJones, Bobblehead, Bobbobagan, Bobianite, Bobo192, Bobrayner, Boffin, Bomac, Bonazzi, Bongwarrior, Bonus bon, Bookandcoffee, Booksacool1, Boothy443, Bovineone, Bowei Huang, BrandonC, Braxton McBragg Burger, Brendan Moody, Brendenhull, Brett Sandford, BrettAllen, Brian, Brian0918, BrianHansen, Brianga, Brianne.hadley, Brichcja, Brighterorange, Bringmemybow, Brotown, Bryan Derksen, Bryce, Brydie1, Bua333, Buaidh, Buchanan-Hermit, Buck49, Bushcarrot, Bwhack, Bwilkins, C mon, C777, CJ, CJLL Wright, CMW275, CSumit, CWenger, Cactus.man, Caerwine, Cakepiejuice, CalJW, Caltas, Cam24, CambridgeBayWeather, Cameron Dewe, Camillaschippa, Can't sleep, clown will eat me, Canadian-Bacon, CanadianLinuxUser, Canderson7, Caniago, Cantabrian, Cantus, Capcom1, CapitalR, Carangoo, Carey Evans, Carinemily, Carnildo, Carrera88, CarterL2011, Casper2k3, Cassivs, Catgut, Cazador, Cedars, Celtic Harper, Centraltendency 01, Cerlian, Ch33syt, Chamdarae, Charlieabeling, CharlotteWebb, Chaz1dave, Che090572, Chevan, Chicken2057, Chillum, Chimesmonster, Chipmunkdavis, Choij, Chris 73, Chris G, Chris Murphy, Chris the speller, ChrisGlew, Chrism, Christiaan, Christo632, Chriswiki, Chun-hian, Chuq, Chuqing.sha, CiTrusD, CipherPixel, City of Destruction, Cjewell, Ckatz, Clam0p, ClamDip, ClaudeMuncey, Clayoquot, Closedmouth, Cnhanxiao, Cocosmooth, Code Cracker 2, Coffee, Collieuk, Colonial from the

Image Sources, Licenses and Contributors

Image:rutan.long-EZ.g-wily.arp.jpg *Source*: http://en.wikipedia.org/w/index.php?title=File:Rutan.long-EZ.g-wily.arp.jpg *License*: unknown *Contributors*: Arpingstone, Gomera-b, Henristosch, Mattes, PeterWD, RosarioVanTulpe

Image:StittsSA-3APlayboyCF-RAD.jpg *Source*: http://en.wikipedia.org/w/index.php?title=File:StittsSA-3APlayboyCF-RAD.jpg *License*: unknown *Contributors*: Original uploader was Ahunt at en.wikipedia

File:Questair 20 Venture N94Y.JPG *Source*: http://en.wikipedia.org/w/index.php?title=File:Questair_20_Venture_N94Y.JPG *License*: unknown *Contributors*: User:Ahunt

File:Swearingen SX-300.jpg *Source*: http://en.wikipedia.org/w/index.php?title=File:Swearingen_SX-300.jpg *License*: unknown *Contributors*: Me

Image:BowersFlyBabyCF-DSH.JPG *Source*: http://en.wikipedia.org/w/index.php?title=File:BowersFlyBabyCF-DSH.JPG *License*: unknown *Contributors*: Ahunt

Image:PietenpolAirCamperUncovered.JPG *Source*: http://en.wikipedia.org/w/index.php?title=File:PietenpolAirCamperUncovered.JPG *License*: unknown *Contributors*: Ahunt, Liftarn

Image:genav.vansrv4.arp.750pix.jpg *Source*: http://en.wikipedia.org/w/index.php?title=File:Genav.vansrv4.arp.750pix.jpg *License*: unknown *Contributors*: Original uploader was Arpingstone at en.wikipedia

Image:MurphyMooseUnderConstruction.JPG *Source*: http://en.wikipedia.org/w/index.php?title=File:MurphyMooseUnderConstruction.JPG *License*: unknown *Contributors*: Original uploader was Ahunt at en.wikipedia

Image:rutan quickie q2.jpg *Source*: http://en.wikipedia.org/w/index.php?title=File:Rutan_quickie_q2.jpg *License*: unknown *Contributors*: Original uploader was Ericg at en.wikipedia

Image:CirrusVK-30N94CM01.jpg *Source*: http://en.wikipedia.org/w/index.php?title=File:CirrusVK-30N94CM01.jpg *License*: unknown *Contributors*: Original uploader was Ahunt at en.wikipedia

File:Gulfstream_V_NASA.jpg *Source*: http://en.wikipedia.org/w/index.php?title=File:Gulfstream_V_NASA.jpg *License*: unknown *Contributors*: user:Stahlkocher

File:First flight2.jpg *Source*: http://en.wikipedia.org/w/index.php?title=File:First_flight2.jpg *License*: unknown *Contributors*: John T. Daniels

Image:Hindenburg at lakehurst.jpg *Source*: http://en.wikipedia.org/w/index.php?title=File:Hindenburg_at_lakehurst.jpg *License*: unknown *Contributors*: U.S. Department of the Navy. Bureau of Aeronautics. Naval Aircraft Factory, Philadelphia, Pennsylvania (USA).

Image:Helios cthomas.jpg *Source*: http://en.wikipedia.org/w/index.php?title=File:Helios_cthomas.jpg *License*: unknown *Contributors*: NASA

Image:Nwa a330-300 n805nw arp.jpg *Source*: http://en.wikipedia.org/w/index.php?title=File:Nwa_a330-300_n805nw_arp.jpg *License*: unknown *Contributors*: Arpingstone, J o, My name

Image:cessna.120.g-btbw.arp.jpg *Source*: http://en.wikipedia.org/w/index.php?title=File:Cessna.120.g-btbw.arp.jpg *License*: unknown *Contributors*: Arpingstone, Gomera-b, JePe, Joshbaumgartner

Image:ixess hang glider arp.jpg *Source*: http://en.wikipedia.org/w/index.php?title=File:Ixess_hang_glider_arp.jpg *License*: unknown *Contributors*: Arpingstone, PeterWD

Image:Lockheed SR-71 Blackbird.jpg *Source*: http://en.wikipedia.org/w/index.php?title=File:Lockheed_SR-71_Blackbird.jpg *License*: unknown *Contributors*: USAF/Judson Brohmer

Image:Towers Schiphol small.jpg *Source*: http://en.wikipedia.org/w/index.php?title=File:Towers_Schiphol_small.jpg *License*: unknown *Contributors*: Ilse@, JMPerez, Kw0, Mattes, Paulbe, Ronaldino

Image:Contrails.jpg *Source*: http://en.wikipedia.org/w/index.php?title=File:Contrails.jpg *License*: unknown *Contributors*: Saperaud, Solon, 1 anonymous edits

File:air.show.utterly.arp.500pix.jpg *Source*: http://en.wikipedia.org/w/index.php?title=File:Air.show.utterly.arp.500pix.jpg *License*: unknown *Contributors*: Arpingstone, Benchill, Dammit, Hohum, KTo288, PlusMinus, Sandstein, 1 anonymous edits

File:Aviatiker-Woche Reims 1909.jpg *Source*: http://en.wikipedia.org/w/index.php?title=File:Aviatiker-Woche_Reims_1909.jpg *License*: unknown *Contributors*: Bin im Garten, LGR109, Maximilian Schönherr, PeterWD

File:Airshow Nellis AFB.JPG *Source*: http://en.wikipedia.org/w/index.php?title=File:Airshow_Nellis_AFB.JPG *License*: unknown *Contributors*: Trevor MacInnis, 1 anonymous edits

File:Patrouille Suisse at ILA 2010 07.jpg *Source*: http://en.wikipedia.org/w/index.php?title=File:Patrouille_Suisse_at_ILA_2010_07.jpg *License*: unknown *Contributors*: User:MatthiasKabel

File:red.bull.air.race.arp.750pix.jpg *Source*: http://en.wikipedia.org/w/index.php?title=File:Red.bull.air.race.arp.750pix.jpg *License*: unknown *Contributors*: Ardfern, Morio, Mtaylor848, Snorky

File:F-111-Fuel-Dump,-Avalon,-VIC-23.03.2007.jpg *Source*: http://en.wikipedia.org/w/index.php?title=File:F-111-Fuel-Dump,-Avalon,-VIC-23.03.2007.jpg *License*: unknown *Contributors*: User:Jjron

File:crash.arp.600pix.jpg *Source*: http://en.wikipedia.org/w/index.php?title=File:Crash.arp.600pix.jpg *License*: unknown *Contributors*: U.S. Air Force photo by Staff Sgt. Bennie J. Davis III

file:New Zealand location map.svg *Source*: http://en.wikipedia.org/w/index.php?title=File:New_Zealand_location_map.svg *License*: unknown *Contributors*: User:NordNordWest

File:Disc Plain red.svg *Source*: http://en.wikipedia.org/w/index.php?title=File:Disc_Plain_red.svg *License*: unknown *Contributors*: User:Amada44

Image:Airport overhead.jpg *Source*: http://en.wikipedia.org/w/index.php?title=File:Airport_overhead.jpg *License*: unknown *Contributors*: Clindberg, Cnyborg, Geofrog, Ingolfson, Maksim, Nilfanion, Ronaldino

File:Flag of New Zealand.svg *Source*: http://en.wikipedia.org/w/index.php?title=File:Flag_of_New_Zealand.svg *License*: unknown *Contributors*: Adambro, Arria Belli, Avenue, Bawolff, Bjankuloski06en, ButterStick, Denelson83, Donk, Duduziq, EugeneZelenko, Fred J, Fry1989, Hugh Jass, Ibagli, Jusjih, Klemen Kocjancic, Mamndassan, Mattes, Nightstallion, O, Peeperman, Poromiami, Reisio, Rfc1394, Shizhao, Tabasco, Transparent Blue, Väsk, Xufanc, Zscout370, 35 anonymous edits

Image:MatamataHobbiton.JPG *Source*: http://en.wikipedia.org/w/index.php?title=File:MatamataHobbiton.JPG *License*: unknown *Contributors*: User:Valley2city

Image:MatamataCows.jpg *Source*: http://en.wikipedia.org/w/index.php?title=File:MatamataCows.jpg *License*: unknown *Contributors*: Gadfium, Lt, XLerate

File:Mt Maunganui 2006.jpg *Source*: http://en.wikipedia.org/w/index.php?title=File:Mt_Maunganui_2006.jpg *License*: unknown *Contributors*: by Shellzzy

File:Loudspeaker.svg *Source*: http://en.wikipedia.org/w/index.php?title=File:Loudspeaker.svg *License*: unknown *Contributors*: Bayo, Gmaxwell, Husky, Iamunknown, Mirithing, Myself488, Nethac DIU, Omegatron, Rocket000, The Evil IP address, Wouterhagens, 18 anonymous edits

File:Mt Maunganui & Tauranga 2006.jpg *Source*: http://en.wikipedia.org/w/index.php?title=File:Mt_Maunganui_&_Tauranga_2006.jpg *License*: unknown *Contributors*: Ingolfson, Shellzzy

File:Mount_Maunganui_main_beach_in_Tauranga,_New_Zealand.jpg *Source*: http://en.wikipedia.org/w/index.php?title=File:Mount_Maunganui_main_beach_in_Tauranga,_New_Zealand.jpg *License*: unknown *Contributors*: User:Pumpmeup

File:Abaconda Tauranga-Boat Sunrise.jpg *Source*: http://en.wikipedia.org/w/index.php?title=File:Abaconda_Tauranga-Boat_Sunrise.jpg *License*: unknown *Contributors*: User:Ed Kruger

File:Abaconda_ocean-2.jpg *Source*: http://en.wikipedia.org/w/index.php?title=File:Abaconda_ocean-2.jpg *License*: unknown *Contributors*: User:QFSE Media

File:Mt Maunganui Beach 2006.jpg *Source*: http://en.wikipedia.org/w/index.php?title=File:Mt_Maunganui_Beach_2006.jpg *License*: unknown *Contributors*: Jase.Lee, QFSE Media, 4 anonymous edits

File:Abaconda park-1.jpg *Source*: http://en.wikipedia.org/w/index.php?title=File:Abaconda_park-1.jpg *License*: unknown *Contributors*: User:QFSE Media

File:Flag of Japan.svg *Source*: http://en.wikipedia.org/w/index.php?title=File:Flag_of_Japan.svg *License*: unknown *Contributors*: Anomie

File:Flag of the People's Republic of China.svg *Source*: http://en.wikipedia.org/w/index.php?title=File:Flag_of_the_People's_Republic_of_China.svg *License*: unknown *Contributors*: User:Denelson83, User:SKopp, User:Shizhao, User:Zscout370

File:Coat of Arms of New Zealand.svg *Source*: http://en.wikipedia.org/w/index.php?title=File:Coat_of_Arms_of_New_Zealand.svg *License*: unknown *Contributors*: User:Sodacan

File:NZL_orthographic_NaturalEarth.svg *Source*: http://en.wikipedia.org/w/index.php?title=File:NZL_orthographic_NaturalEarth.svg *License*: unknown *Contributors*: User:Gringer

File:Increase2.svg *Source*: http://en.wikipedia.org/w/index.php?title=File:Increase2.svg *License*: unknown *Contributors*: Sarang

File:Detail of 1657 map Polus Antarcticus by Jan Janssonius, showing Nova Zeelandia.png *Source*: http://en.wikipedia.org/w/index.php?title=File:Detail_of_1657_map_Polus_Antarcticus_by_Jan_Janssonius,_showing_Nova_Zeelandia.png *License*: unknown *Contributors*: Avenue

File:Polynesian Migration.svg *Source*: http://en.wikipedia.org/w/index.php?title=File:Polynesian_Migration.svg *License*: unknown *Contributors*: User:Gringer

File:Treatyofwaitangi.jpg *Source*: http://en.wikipedia.org/w/index.php?title=File:Treatyofwaitangi.jpg *License*: unknown *Contributors*: William Hobson, James Freeman, and James Busby (English version); Henry Williams and Edward Williams (Māori translation)

File:John Key National Party2.jpg *Source*: http://en.wikipedia.org/w/index.php?title=File:John_Key_National_Party2.jpg *License*: unknown *Contributors*: Guo's

Image:Elizabeth II greets NASA GSFC employees, May 8, 2007 edit.jpg *Source*: http://en.wikipedia.org/w/index.php?title=File:Elizabeth_II_greets_NASA_GSFC_employees,_May_8,_2007_edit.jpg *License*: unknown *Contributors*: NASA/Bill Ingalls

Image:Jerry Mateparae 090529-N-8623G-003.jpg *Source*: http://en.wikipedia.org/w/index.php?title=File:Jerry_Mateparae_090529-N-8623G-003.jpg *License*: unknown *Contributors*: USDOD MC2 Elisia Gonzales

File:Bowen House Beehive Parliament.JPG *Source*: http://en.wikipedia.org/w/index.php?title=File:Bowen_House_Beehive_Parliament.JPG *License*: unknown *Contributors*: Original uploader was Midnighttonight at en.wikipedia

File:E 003261 E Maoris in North Africa July 1941.jpg *Source*: http://en.wikipedia.org/w/index.php?title=File:E_003261_E_Maoris_in_North_Africa_July_1941.jpg *License*: unknown *Contributors*: Unidentified New Zealand official photographer

File:New Zealand trench Flers September 1916.jpg *Source*: http://en.wikipedia.org/w/index.php?title=File:New_Zealand_trench_Flers_September_1916.jpg *License*: unknown *Contributors*: Royal Engineers No 1 Printing Company

File:Realm of New Zealand-2.PNG *Source*: http://en.wikipedia.org/w/index.php?title=File:Realm_of_New_Zealand-2.PNG *License*: unknown *Contributors*: Realm_of_New_Zealand.png: derivative work: Sesmith

File:commons-logo.svg *Source*: http://en.wikipedia.org/w/index.php?title=File:Commons-logo.svg *License*: unknown *Contributors*: Anomie

File:New Zealand 23 October 2002.jpg *Source*: http://en.wikipedia.org/w/index.php?title=File:New_Zealand_23_October_2002.jpg *License*: unknown *Contributors*: Jacques Descloitres, MODIS Rapid Response Team, NASA/GSFC

File:AbelTasmanNP.jpg *Source*: http://en.wikipedia.org/w/index.php?title=File:AbelTasmanNP.jpg *License*: unknown *Contributors*: Arria Belli, Ingolfson, Steffen84

File:TeTuatahianui.jpg *Source*: http://en.wikipedia.org/w/index.php?title=File:TeTuatahianui.jpg *License*: unknown *Contributors*: Apalsola, Ingolfson, Kahuroa, Tony Wills, Winterkind

File:Tararua Range and Mt Dundas 20 August 1909 edit.jpg *Source*: http://en.wikipedia.org/w/index.php?title=File:Tararua_Range_and_Mt_Dundas_20_August_1909_edit.jpg *License*: unknown *Contributors*: Tararua_Range_and_Mt_Dundas_20_August_1909.jpg: Adkin, George Leslie. Collection in Alexander Turnbull Library: Photographs of New Zealand geology, geography, and the Maori history of Horowhenua, PA1-f-008-003

File:MilfordSound.jpg *Source*: http://en.wikipedia.org/w/index.php?title=File:MilfordSound.jpg *License*: unknown *Contributors*: Original uploader was Wikikiwiman at en.wikipedia

File:Romney Ewe and Lamb.jpg *Source*: http://en.wikipedia.org/w/index.php?title=File:Romney_Ewe_and_Lamb.jpg *License*: unknown *Contributors*: Black Angus Girl, Fneep, Shirt58, Steven Walling, 3 anonymous edits

File:New Zealand population over time - small.png *Source*: http://en.wikipedia.org/w/index.php?title=File:New_Zealand_population_over_time_-_small.png *License*: unknown *Contributors*: User:Avenue

File:Lion dancers at the Auckland lantern festival 2010.jpg *Source*: http://en.wikipedia.org/w/index.php?title=File:Lion_dancers_at_the_Auckland_lantern_festival_2010.jpg *License*: unknown *Contributors*: User:Avenue

File:Ratana Church Raetihi.jpg *Source*: http://en.wikipedia.org/w/index.php?title=File:Ratana_Church_Raetihi.jpg *License*: unknown *Contributors*: Alan Liefting, Arria Belli, Avenue, Diego Grez, Ingolfson

File:KupeWheke.jpg *Source*: http://en.wikipedia.org/w/index.php?title=File:KupeWheke.jpg *License*: unknown *Contributors*: User:Kahuroa

File:Cook Islands dancers at Auckland's Pacifica festival 3a.jpg *Source*: http://en.wikipedia.org/w/index.php?title=File:Cook_Islands_dancers_at_Auckland's_Pacifica_festival_3a.jpg *License*: unknown *Contributors*: User:Avenue

File:Hinepare.jpg *Source*: http://en.wikipedia.org/w/index.php?title=File:Hinepare.jpg *License*: unknown *Contributors*: Lindauer, Gottfried, 1839-1926

File:Hillary statue and Mount Cook.jpg *Source*: http://en.wikipedia.org/w/index.php?title=File:Hillary_statue_and_Mount_Cook.jpg *License*: unknown *Contributors*: Jonathan Keelty from Hong Kong, Hong Kong SAR

Printed by Books on Demand GmbH, Norderstedt / Germany